Christoph M. Bayer

Characterization of High Temperature Superconductor Cables for Magnet Toroidal Field Coils of the DEMO Fusion Power Plant

Karlsruher Schriftenreihe zur Supraleitung Band 018

HERAUSGEBER

Prof. Dr.-Ing. M. Noe
Prof. Dr. rer. nat. M. Siegel

Eine Übersicht über alle bisher in dieser Schriftenreihe erschienene Bände finden Sie am Ende des Buches.

Characterization of High Temperature Superconductor Cables for Magnet Toroidal Field Coils of the DEMO Fusion Power Plant

by
Christoph M. Bayer

KIT Scientific Publishing

Dissertation, Karlsruher Institut für Technologie (KIT)
Fakultät für Elektrotechnik und Informationstechnik, 2016
Hauptreferent: Prof. Dr.-Ing. Mathias Noe
Korreferentin: Prof. Dr. Luisa Chiesa, Ph.D., MIT

Impressum

KIT Scientific Publishing

Karlsruher Institut für Technologie (KIT)
KIT Scientific Publishing
Straße am Forum 2
D-76131 Karlsruhe

KIT Scientific Publishing is a registered trademark of Karlsruhe Institute of Technology. Reprint using the book cover is not allowed.

www.ksp.kit.edu

Print on Demand 2017 – Gedruckt auf FSC-zertifiziertem Papier

ISSN 1869-1765
ISBN 978-3-7315-0605-8
DOI 10.5445/KSP/1000062054

Characterization of High Temperature Superconductor Cables for Magnet Toroidal Field Coils of the DEMO Fusion Power Plant

Zur Erlangung des akademischen Grades eines

DOKTOR-INGENIEURS

von der Fakultät für

Elektrotechnik und Informationstechnik

des Karlsruher Institut für Technologie

genehmigte

DISSERTATION

von

Dipl.-Phys. Christoph M. Bayer

geb. in: Pforzheim

Tag der mündlichen Prüfung: 02.11.2016

Hauptreferent: Prof. Dr.-Ing. Mathias Noe[1]

Korreferentin: Prof. Dr. Luisa Chiesa, Ph.D., MIT[2]

[1]Institut für Technische Physik (ITEP), Karlsruher Institut für Technologie (KIT), Deutschland

[2]Department of Mechanical Engineering, Tufts University, Medford, Massachusetts, USA

Acknowledgements

Firstly, I would like to express my gratitude to my supervisor, Prof. Dr.-Ing. Mathias Noe, for his continuous support and feedback on my proceeding work. I highly appreciated the very effective and constructive meetings and the helpful guidance I was given. Furthermore, it was always a pleasure to participate in many seminars and workshops organised at the Institute for Technical Physics under Prof. Noe's leadership.

I would like to thank my second referee, Prof. Dr. Luisa Chiesa, for her scientific input and her linguistic improvements, that helped a great deal to finish the final version of this thesis.

My special gratitude goes to my advisor Dr. Klaus-Peter Weiss. I would like to thank him for his harmonic group leadership, which leaves room for free development and new ideas, and the great opportunity to frequently participate in international conferences, workshops and seminars, which had a huge impact on my scientific work.

Special thanks go to Christian Lange and Frank Gröner. Many measurements could not have been performed, if not for their outstanding, technical support and their motivation to repair or replace broken or damaged equipment.

I would like to thank Dr.-Ing. Christian Barth, Dr. Nadezda Bagrets, Pattabhi V. Gade, Dr.-Ing. Philipp Krüger, Sascha Westenfelder, Bianca Purr, Elisabeth Weiss, Valentin Tschan and the entire CryoMaK team for many on- and off-topic discussions, social activities and barbecues, which have always been a recreating highlight during a busy week.

Finally, I would like to express my sincere gratitude to my parents, Klaus and Beate, to my brothers, Jan and Matthias, and to my lovely wife, Mariana. I would never have made it this far if not for the support of my family, to whom I shall ever be indebted.

Christoph M. Bayer
24.08.2016

Kurzfassung

Kernfusion gilt als eine der Schlüsseltechnologien um den Grundbedarf an elektrischer Energie nachhaltig zu decken. Die Fusion, als Energiequelle ohne Emission von Treibhausgasen und Produktion radioaktiver Abfälle, kommt der Maxime einer sauberen Energiequelle sehr nahe.

Aktuelle Fusionsexperimente, darunter der *International Thermonuclear Experimental Reactor* (ITER), legen den Fokus auf die wissenschaftliche Untersuchung der Plasmaphysik. Um die Lücke zwischen rein experimentellem Reaktor und ökonomischem Fusionskraftwerk zu schließen, bedarf es eines Demonstrationsreaktors (DEMO), der die wissenschaftlichen und technischen Entwicklungen in industrielle Standards überführt. Offizielle Zeitpläne von EUROfusion sehen derzeitig eine Inbetriebnahme des europäischen DEMO-Reaktors für 2050 vor.

Diese Dissertation befasst sich mit der Untersuchung von Hochtemperatursupraleiterkabeln auf ihre Anwendbarkeit in Toroidalfeldspulen des DEMO-Reaktors. Damit einher gehen die explizite Beschreibung der Messapparatur zur Untersuchung der Kabelproben, die Verarbeitung der Hochtemperatursupraleiter während des Herstellungsprozesses und die Vorbereitung der Kabelproben für die anschließenden Messungen in starken Magnetfeldern bei Lorentzkräften von mehreren Kilonewton.

Diese Arbeit eröffnet mit einer Einführung in die Thematik der Kernfusion und der technischen Hochtemperatursupraleiter. Besondere Gewichtung finden die bisherige ITER Magnettechnologie und die technischen Magnetanforderungen an DEMO.

Der zweite Teil der Arbeit setzt sich mit der Kabelherstellung, insbesondere dem Weichlöten von *RE*BCO-Hochtemperatursupraleitern auseinander. Der Verarbeitungsprozess, das verwendete Weichlot und die Serie des *RE*BCO-Supraleiters haben großen Einfluss auf das erzielte Ergebnis. Verschiedenste Techniken und Ergebnisse wurden erfasst, statistisch evaluiert und abschließend für ihre weitere Anwendung an den Kabelproben bewertet.

Es folgt eine explizite Beschreibung der Messapparatur zur Untersuchung der Kabelproben. Diese Apparatur wurde im Rahmen der Arbeit technisch erweitert, um auch Proben mit besonderes großen Probendurchmessern in Abhängigkeit von Magnetfeld, Temperatur und Zugspannung untersuchen zu können.

Für die eigentliche Untersuchung werden unterschiedlichste Typen an Hochtemperatursupraleiterkabeln herangezogen. Der Probenumfang umfasst zwei *Roebel Assembled Coated Conductors* (RACC), drei *Stacked Tape Cables* (STC), drei *Twisted Stacked Tape Cables* (TSTC) und ein *Cable on Round Core* (CORC). Jedes dieser Kabel wurde bei Temperaturen von 4.2 K in flüssigem Helium und in einem Hintergrundmagnetfeld von bis zu 12 T getestet.

Insbesondere die beiden RACC-Proben bedurften hierzu einer besonderen Stabilisierung gegen mechanische Verformung aufgrund von Lorentzkräften. Diese Stabilisierung setzte die Untersuchung der Widerstandsfähigkeit der RACC-Proben gegen senkrecht ausgeübten Druck voraus. Eine STC-Probe und zwei TSTC-Proben wurden darüber hinaus auf ihre Temperaturabhängigkeit untersucht. Für eine STC-Probe wurde ein irreversibler Zugversuch bei 12 T und 4.2 K durchgeführt.

Abschließend werden vier mögliche Kabelmodelle auf Grundlage der experimentellen Erkenntnisse und theoretischen Anforderungen an aktuelle Toroidalfeldspulen für den DEMO-Reaktor berechnet und verglichen. Sowohl die Materialzusammensetzung als auch die Zugspannung am Supraleiter haben dabei grundlegenden Einfluss auf die finalen Modelle.

Contents

1 Introduction and motivation

According to the key world energy statistics from 2015 [OI15] by the Organisation for Economic Co-operation and Development (OECD) and the International Energy Agency (IEA), the total world energy consumption in 2013 was 13541 Mtoe [OI15], which corresponds to 1.58×10^{14} kWh. The ratio of consumed energy to generate electricity is 18% or 2.85×10^{13} kWh. By-product of the generated energy is the greenhouse gas CO_2. Considering fuel combustion, 38×10^9 tons of CO_2 were emitted in 2013 [OI15]. Future estimations predict a world energy consumption of more than 1.8×10^{14} kWh for 2040 at best [OI15]. More realistic estimations on basis of new policies under consideration, exceed the two petawatt hours barrier and predict a constant CO_2 emission of 38×10^9 tons [OI15].

Such high values demand to consider more environment-friendly energy sources - desirably most efficient, to keep CO_2 to a minimum level. Renewable energies such as wind turbines and solar power stations are volatile and therefore highly dependent on season, climate, weather and location. Conventional power stations on basis of fossil fuels can be operated in steady-state, but then again, with the exception of nuclear fission power stations, are not sustainable. Typical fossil fuels and their corresponding specific energies are listed in table 1.1.

Table 1.1: Typical fossil fuels and their specific energies. Data from [Smi00, Tou16].

fossil fuel	specific energy (MJ/kg)	reaction
coal	32 - 35	combustion
gasoline	43.4	combustion
diesel	40.3	combustion
methane / natural gas	45 - 50	combustion
hydrogen	120	fuel cell / combustion
thorium 232	7.9×10^7	fission
uranium 235	7.9×10^7	fission
tritium + deuterium	2.8×10^8	fusion

It is not the fossil itself, but the physical or chemical reaction to access the stored energy, which decides between an energy output of few megajoules or preferable terajoules per kilogram invested fuel. The highest energy efficiency levels for conventional power stations are obtained with fission reactions. A severe uncertainty for fission is the permanent repository of nuclear waste with half-life between thousands and millions of years. As long as this question is not answered, fission cannot be considered an ecological solution.

In near future, a potential energy source, which could combine high energy efficiency and provide baseload power with environmental sustainability, is controlled nuclear fusion. Its fuel deuterium is virtually unlimited [Cra61], fusion reactions do not produce greenhouse gases, nuclear fusion reactions are controlled reactions which cannot degenerate into uncontrolled chain reactions [Wika, Wikb] and, except for the radioactive device structure which is intended to be fully recyclable after only 100 years, do not produce radioactive nuclear waste.

Today's most important fusion device is ITER, the International Thermonuclear Experimental Reactor. Its first plasma ignition is expected in 2025. Parallel to ITER construction, the European Consortium for the Development of Fusion Energy, EUROfusion, and the European Joint Undertaking for ITER and the Development of Fusion Energy, Fusion for Energy (F4E), motivate a first fusion DEMOnstration Power Plant, DEMO, to be operational in 2040 [Rom12]. Most DEMO designs aim for larger plasma volumes and higher plasma aspect ratios than ITER [FKW+14]. Both requires huge, superconducting magnet coils to confine the plasma inside a toroidal magnetic field of up to 6.8 T [Kem12] along the centre line. Such technical challenges cannot be met with former technological standards, but demand technological innovation.

With the development of second generation (2G) high temperature superconductors (HTS), often referred to as *Rare-Earth*-Barium-Copper-Oxide (*RE*BCO) coated conductors, new possibilities for fusion magnet technology opened up. *RE*BCO coated conductors exhibit a high mechanical tensile strength, an increased temperature margin and a high engineering current density. As *RE*BCO superconductors offer a potential application for fusion relevant magnets, various research groups are presently investigating in different cable concepts for fusion magnet coils [TCBM12, GGP+14, WFB+16, ADC+15, LGH12].

The main focus of this thesis is the investigation and evaluation of present high temperature superconductor cable concepts for their applicability in toroidal field coils of the DEMO power plant according to the DEMO July 2012 design. This includes an explicit description of both measurement facilities to investigate *RE*BCO tapes and sub-size cable samples, the *RE*BCO treatment during the manufacturing process and the preparation of each sub-size cable sample for its in-field measurements.

In chapter 2 fusion technology is introduced and an overview of present fusion experiments and state-of-the-art fusion devices is given.

*RE*BCO high temperature superconductors are introduced in chapter 3, followed by detailed specifications on magnet technology of tokamak reactors. Technological standards for ITER are summarized and design parameters for DEMO on basis of the July 2012 design are explicitly discussed.

Important data on soft-soldering of *RE*BCO superconductors, gathered from many publications and experiments over past years is summarized in chapter 4. As the core process during the manufacturing of *RE*BCO cables, soft-soldering of cable joints and cable terminations plays an important role in the development of high temperature superconductor cables for fusion and non-fusion applications.

A detailed description on the measurement facilities and their upgrade to investigate *RE*BCO single tapes and sub-size cables in background magnetic fields at different temperatures and under tensile load is given in chapter 5. Furthermore, this chapter contains information about the sample preparation and the sample holders for all experiments.

Chapter 6 comprises sample descriptions, sample properties and in-field measurements of all high current HTS sub-size cable concepts investigated within this thesis. The result for each sample test is discussed individually and afterwards put in context to the overall investigation. Special attention is given to the Roebel Assembled Coated Conductor (RACC), which is the only *RE*BCO cable concept with a full strand transposition, as it is used in the Nb_3Sn ITER full-size cable in conduit conductor.

In chapter 7 the experimental results of foregoing sub-size cable investigations are extended by an evaluation of the most promising cable concepts for the DEMO power plant according to the DEMO July 2012 design. Detailed extrapolations to the most promising full-size fusion cables are proposed.

This thesis concludes with chapter 8, a final summary of the main aspects and an outlook on future developments.

2 Fusion technology

Ever since Hans Albrecht Bethe claimed fusion to be the power of stars in 1939 [Bet39], on behalf of which he later received the Nobel price in physics, the idea of controlled fusion spread world wide. In 1946, George Paget Thomson and Moses Blackman patented the Z-pinch technique, where electrical current in a plasma induces a magnetic confinement which leads to a plasma pinch [TB46]. Only four years later, the soviet scientists Andrei Dmitrijewitsch Sakharov and Igor Jewgenjewitsch Tamm suggest the токамак (toroidal chamber with magnetic coils), the first device to confine plasma inside a magnet torus [Azi12]. Almost at the same time, in 1951, Lyman Spitzer proposes the stellarator, an alternative device to the tokamak, which shares the idea of confining plasma inside a magnet torus. A decade later, in 1962, the first tokamak device T3 was put into operation at Kurchatov Institute, Moscow [Deu11]. Peak temperatures of up to 1 keV were measured and published in 1969 [PRF⁺69]. In the following years, various tokamak experimental reactors were developed. Since 1983, the Joint European Torus (JET) with the largest magnetic plasma confinement is operational at Culham, United Kingdom [RG84]. In 1985, first plasma is reported at the Japan Torus JT-60 at Naka, Japan [AAA⁺86]. Tore Supra at Cadarache, France, and the Axially Symmetric Divertor Experiment (ASDEX) Upgrade, Garching, Germany follow with first plasma in 1988 [GEG⁺90] respectively 1991 [Deu11]. Parallel to the tokamak development, the first stellarators are put into operation. The Wendelstein 7-AS experiment at Garching, Germany, starts in 1988 [HBB⁺08], followed by the Large Helical Device (LHD) at Toki, Japan, in 1998 and the Helically Symmetric Experiment (HSX) at Madison, USA, in 1999 [Deu11]. While JET demonstrates a maximum fusion power of 16 MW [KT98], Tore Supra holds, with a duration of more than six minutes, the record for longest plasma period achieved [HMB⁺04]. More than 200 tokamak experiments world wide have been performed up to now. A collection on spherical and conventional tokamak devices and their properties can be found at [Bal15].

Within few decades, tokamak and stellarator became the most popular fusion devices. Today's most prominent and largest tokamak is the International Thermonuclear Experimental Reactor (ITER), with 35 countries contributing to its construction at Cadarache, France. It is handled as the predecessor of the first DEMOnstration power plant (DEMO). The most prominent stellarator, Wendelstein 7-X, at Greifswald, Germany, is operational since 2015 [OKWZ16]. In the following chapter, stellarator and tokamak are introduced with main focus on the magnetic plasma confinement.

2.1 Nuclear fusion and plasma confinement

Natural nuclear fusion reactions are stellar reactions. A star's gravitation is the optimal plasma confinement, which provides sufficient hydrostatic pressure to keep the plasma ignition alive. In light stars, such as our sun, the dominant process is the proton-proton chain reaction

$$\begin{aligned} {}^{1}\mathrm{H} + {}^{1}\mathrm{H} &\rightarrow {}^{2}H + p^{+} + \nu && (0.42\,\mathrm{MeV}) \\ {}^{1}\mathrm{H} + {}^{2}\mathrm{H} &\rightarrow {}^{3}\mathrm{H} + \gamma && (5.49\,\mathrm{MeV}) \\ {}^{3}\mathrm{H} + {}^{3}\mathrm{H} &\rightarrow {}^{4}\mathrm{He} + {}^{1}\mathrm{H} + {}^{1}\mathrm{H} && (12.86\,\mathrm{MeV}). \end{aligned}$$

The solar temperature of more than 10^7 K and the hydrostatic pressure of up to 10^{16} Pa provide sufficient energy to increase the probability for quantum tunnelling and overcome the coulomb barrier of the fusing nuclei, visualized in figure 2.1.

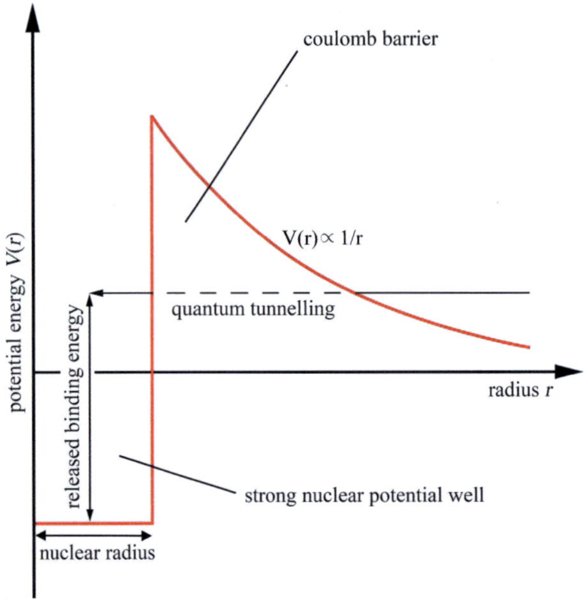

Figure 2.1: With increasing particle energy the probability for quantum tunnelling through the coulomb barrier increases. After penetrating the coulomb barrier, both nuclei are affected by the strong force and fuse to a larger nucleus. The difference between the binding energy of the separated and the combined nuclei is released.

As maximum temperature and hydrostatic pressure of a star scale with its mass, larger stars are even able to fuse heavier elements, such as carbon, nitrogen and oxygen. However, any natural fusion chain reaction is limited by the binding energy of the reacting nuclei. Iron and nickel have the highest binding energies and therefore represent the upper limit of stellar fusion reactions. The binding energy, depending on the number of nucleons, is shown in figure 2.2.

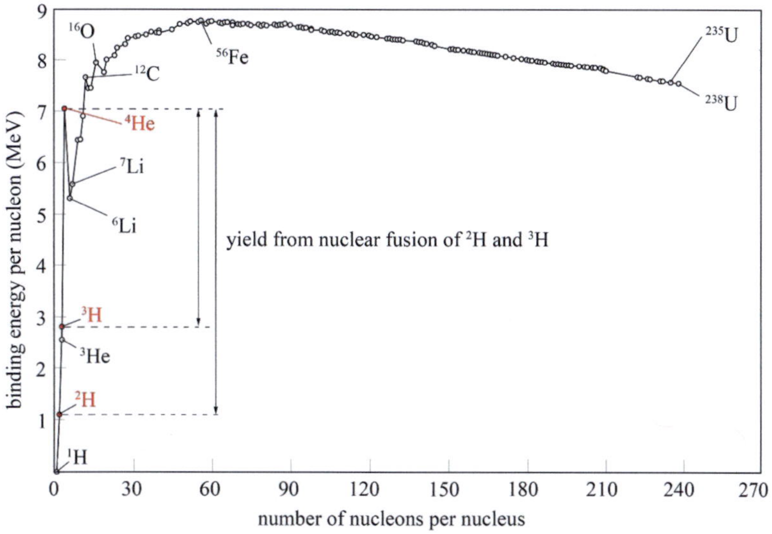

Figure 2.2: Binding energy depending on the number of nucleons. For iron and nickel the binding energy is maximal. Compared to any other hydrogen fusion reaction, the deuterium-tritium reaction is the reaction with the highest energy output and the lowest requirement to reach the Lawson criterion on earth.

Although solar fusion is well understood, it is not feasible to try to adopt solar fusion reactions to terrestrial applications. The boundary conditions for solar fusion are too difficult to obtain.

Requirement for any fusion reaction to reach the point of ignition is to satisfy the Lawson criterion [Law57]. For solar proton-proton reactions, the Lawson criterion is primarily reached by the enormous, solar hydrostatic pressure. For terrestrial fusion reactions, where the plasma is confined inside magnetic fields, such pressure cannot be reached on large scale. However, other fusion reactions have lower requirements, such as the deuterium-tritium reaction

$$^2\mathrm{H} + ^3\mathrm{H} \rightarrow ^4\mathrm{He} + \mathrm{n}\,(17.6\,\mathrm{MeV}),$$

which is therefore the optimum candidate for terrestrials fusion applications. Its Lawson criterion can already be reached for low magneto-hydrostatic pressure and plasma temperatures of 10^8 K, its cross-section is much higher than the cross-section of the proton-proton reactions and its energy yield of 17.6 MeV is very high.

First plasma experiments induced the magnet confinement by a current along a cylindrical plasma, expressed by Ampère's circuital law

$$\nabla \times \overrightarrow{B} = \mu_0 \cdot \overrightarrow{j}.$$

The resulting magnet force $\overrightarrow{j} \times \overrightarrow{B}$ of the induced field and the plasma current pinches the plasma and increases the plasma pressure. Unfortunately, such z-pinches are highly unstable. In order eliminate pinch instabilities it is necessary to add a magnet component in z-direction of the plasma axis. This is best achieved by confining the plasma inside a helical magnetic field, as shown in figure 2.3. Additional bending of the helical field into a torus form, defined by the toroidal radius R_0 and the minor radius a, allows for a closed plasma circuit without the necessity of magnetic mirrors. The helicity of such toroidal magnetic field is defined by the safety-factor q, which is the number of toroidal transits required to circle one poloidal transit. Although the bending of the helical field in figure 2.3 would lead to a symmetric torus, torus symmetry is no requirement for a consistent plasma confinement. While a tokamak generates a rotationally symmetric plasma torus, a stellarator benefits from its asymmetric plasma confinement. Yet, both designs are far from being fully developed and each design has still to face its individual technological challenges.

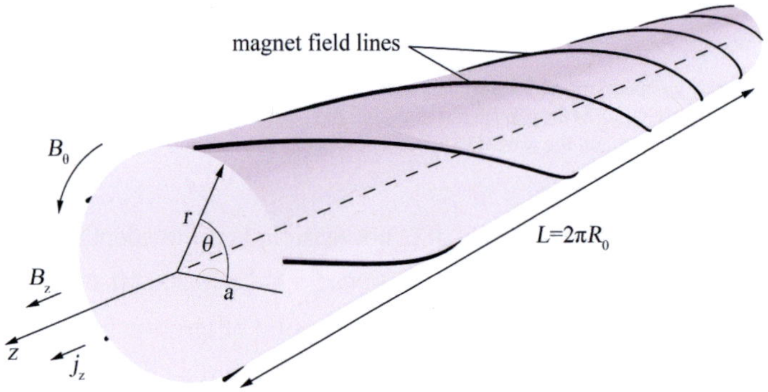

Figure 2.3: Helical magnetic confinement as superposition of the poloidal field B_θ and the longitudinal field B_z.

2.2 Stellarator

In a stellarator (Latin: *stella* = star), the poloidal field B_θ is not induced by a plasma current as in a z-pinch or tokamak, but by various external magnet coils. In the optimum case, the sum of all coil currents in toroidal direction is zero. Hence, no plasma current is induced. The typical multipole magnetic field of a stellarator is shown in figure 2.4. Three types of stellarators are distinguished.

- Classic stellarator: the classic stellarator induces its magnetic field by n pairs of helical conductors with opposite currents which are surrounded by toroidal field coils, that share the same current direction. As the sum of all toroidal currents is zero, no plasma current is induced. Figure 2.5 visualizes the magnetic field and the resulting plasma cross-section of a classic stellarator. Wendelstein 7-A, operated from 1975 to 1985, was one of the first devices.
- Torsatron/Heliotron: the torsatron, also known as heliotron, consists of n helical conductors that share the same current direction. Additional planar coils are required to neutralize the vertical component of the helical field. As the sum of all toroidal currents is different from zero, a plasma current is induced. Figure 2.5 visualizes the magnetic field and the resulting plasma cross-section of a torsatron. The most famous torsatron is the Large Helical Device (LHD) in Japan.
- Modular stellarator: Modules of non-planar magnet coils are connected in series to induce the multi-polar magnetic field. Each magnet coil design is calculated numerically and optimized in various iterations, which allows for small coil radii and low operating currents. As each coil has a closed current loop, the sum of all currents in toroidal direction is zero and a plasma current is not induced. Figure 2.5 visualizes the magnetic field and the resulting plasma cross-section of a modular stellarator. The largest and only optimised modular stellarator is Wendelstein 7-X (W7-X) 2.2.

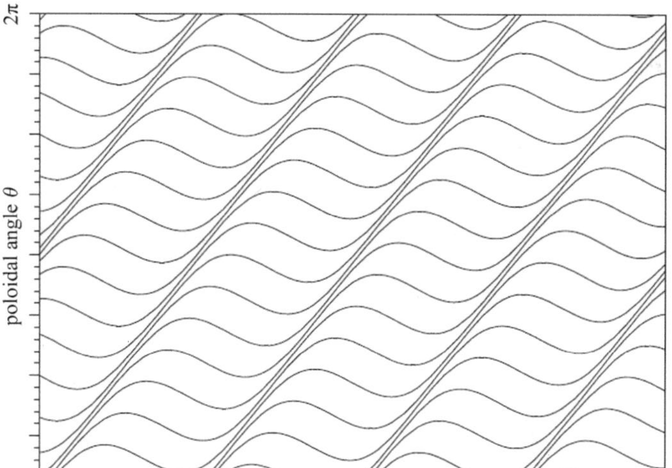

Figure 2.4: Multipole field of a stellarator. Picture from [WWZ10].

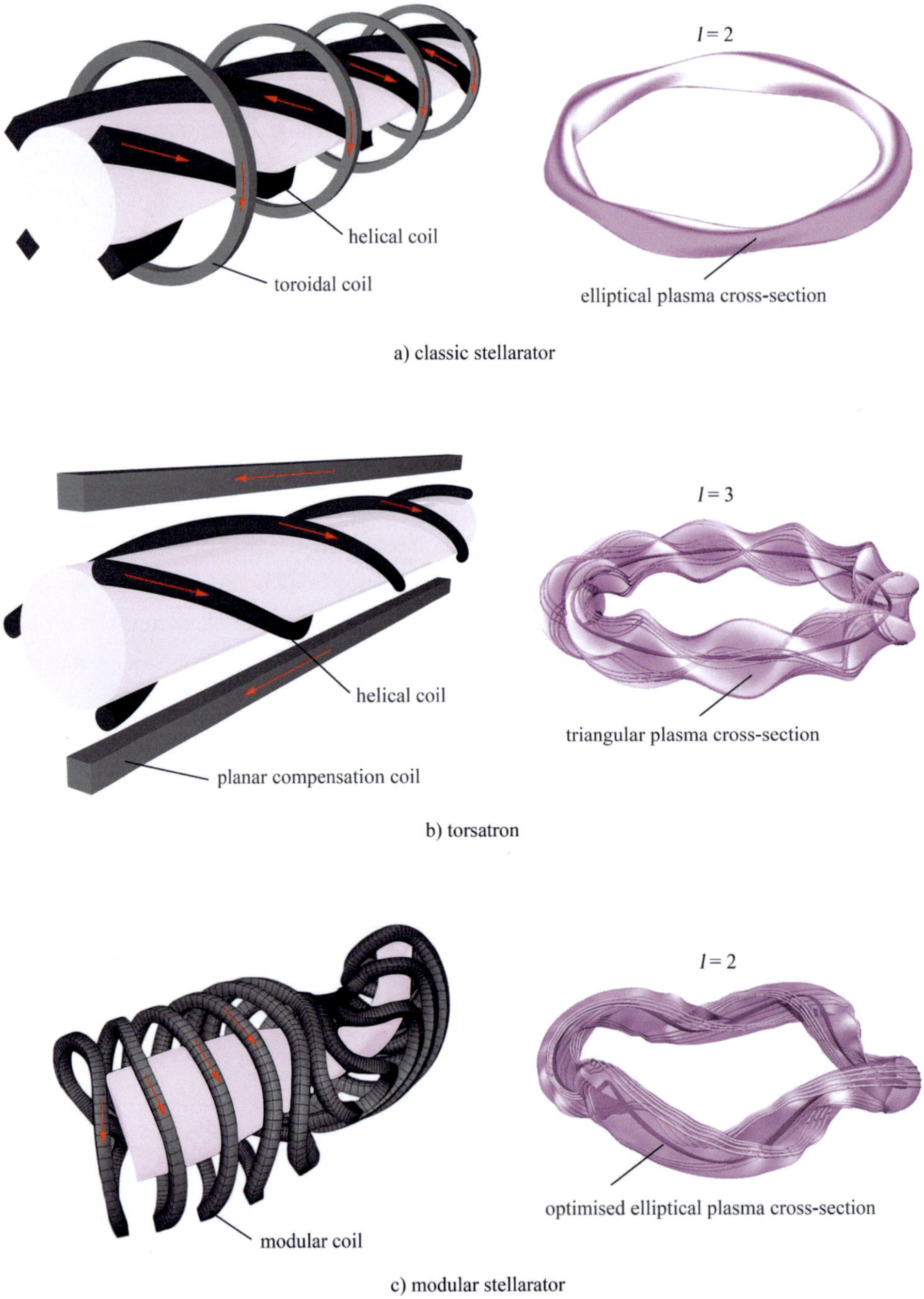

Figure 2.5: Magnet coils of the three different stellarator designs with the corresponding l-times folded magnetic field. Red arrows mark the current direction. Original plasma shape of the torsatron and the modular stellarator are taken from [Spo15].

Although the calculation of the complicated multipole magnetic field is dependent on supercomputers and the magnet structure requires the development and construction of five different, superconducting toroidal confinement coils with a uniform, bendable and highly stable superconductor cable, the modular stellarator has two advantages over the tokamak design:

- As the poloidal field is not induced by a plasma current, but by external magnet coils, the stellarator can be operated in steady-state mode. For a fusion power station this could mean continuous power generation over the regular time of operation without the urge to consider microwave heating or neutral beam injection to drive the plasma current, as it is required for a non-pulsed, steady-state operated tokamak.
- Plasma currents often go along with a high energetic plasma instability: the plasma disruptions. The kinetic energy of the plasma is transferred to the blanket and the collapsing poloidal field induces eddy currents in the vacuum vessel, which again interact with the remaining magnetic field. This stresses the mechanical structure and can even result in damage or displacement of the vacuum vessel (vertical displacement events). A classic or modular stellarator without a net plasma current does not suffer from plasma disruptions.

Wendelstein 7-X (W7-X)

Wendelstein 7-X (W7-X) is the most modern and only modular stellarator with an optimised magnetic field, which started operation in 2015 [OKWZ16]. It is designed to test the new, optimised stellarator plasma confinement with modular coils and to reach the same plasma stability as it is provided by a tokamak of the same size.

The magnet system of W7-X is composed of five equivalent magnet modules. Each module consists of five different pairs of superconducting, non-planar coils, which together form the double-folded, optimized, magnetic plasma confinement. Twenty additional superconducting, planar magnet coils allow for corrections and modulations of the toroidal field. Additional non-superconducting coils are used behind the divertors and outside the plasma vessel, to adjust the magnetic field and compensate unexpected field deviations. A model of the assembled Wendelstein 7-X experiment is shown in figure 2.6. All five equivalent coil modules together form the pentagon shaped magnet torus.

Compared to other tokamak experiments, W7-X is relatively small. The precise definition of the plasma form allows for smaller minor plasma radii. However, to obtain larger plasma volumes, large major plasma radii are required. For a stellarator with the same thermal energy as ITER, its major plasma radius $R_{0-\mathrm{HELIAS}}$ would be twice the ITER major plasma radius $R_{0-\mathrm{ITER}}$.

The most important facts on W7-X are summarized and compared to the ITER and DEMO experiments in table 2.1 in subsection 2.3.

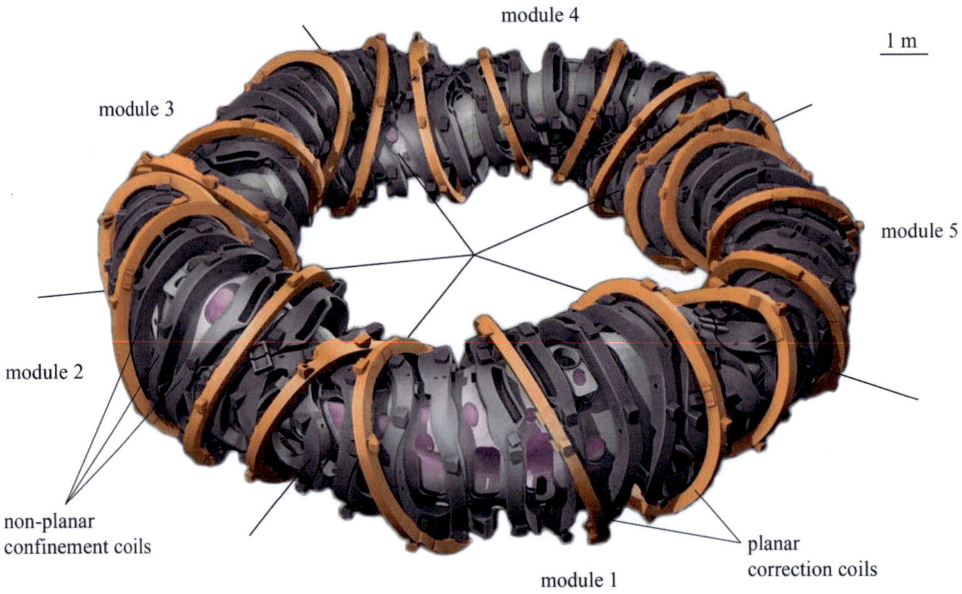

Figure 2.6: Modular magnet structure of Wendelstein 7-X with 50 non-planar confinement- and 20 planar correction coils. Original picture from [Bra].

2.3 Tokamak

As the tokamak concept was already invented in 1950 and has been tested in hundreds of experiments ever since, it is better understood than the stellarator. With the International Thermonuclear Experimental Reactor (ITER) under construction and a European Demonstration Power Plant (DEMO) already in development, the tokamak technology is only few, if not one experimental reactor from a first commercial fusion power plant.

The plasma confinement of a tokamak is originally based on the z-pinch, where a plasma current induces a poloidal, magnetic field. However, in order to avoid z-pinches, the plasma current is guided inside a magnetic torus, which is induced by external toroidal field coils. Superposition of the magnetic poloidal and toroidal field yields in a magnetic torus with helically wound field lines. Figure 2.7 demonstrates the induction of the helical, magnetic plasma confinement.

Due to the torus shape of the plasma it is not feasible to drive the plasma current by a potential gradient as in a pinch machine. Instead, the plasma current is induced by additional, poloidal

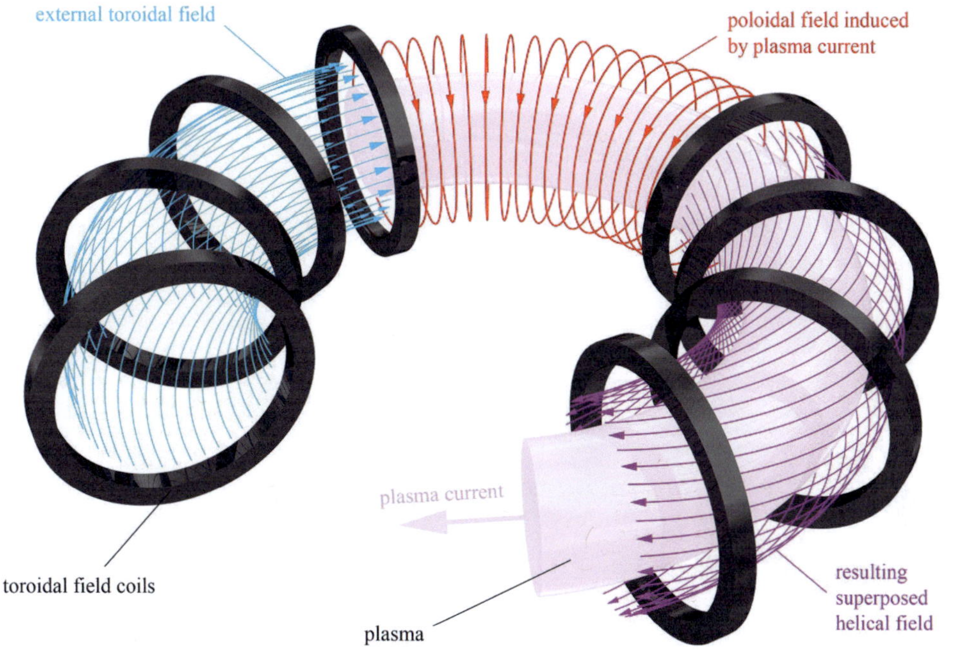

Figure 2.7: Magnetic plasma confinement of a tokamak. The helically wound, magnetic torus yields from superposition of toroidal and poloidal magnetic fields.

field coils. These include a solenoid inside the torus centre and various coils outside the plasma torus. With each linear ramp of the poloidal field coils, a constant plasma current is induced according to the electromagnetic induction

$$\nabla \times \overrightarrow{E} = -\frac{\partial \overrightarrow{B}}{\partial t}.$$

After maximum field is achieved, the poloidal field coils are ramped in opposite direction, which results in an opposite plasma current. Due to its cyclic operation the original tokamak is no steady-state fusion device. In order to achieve steady-state operation, various alternatives to drive the plasma current are under investigation. Neutral beam injection (NBI) of high energetic deuterium and tritium particles (~1 MeV) and wave induced electron cyclotron current drive (ECCD), combined with occurring bootstrap currents due to an inhomogeneous plasma density, could be a possible solution to overcome the pulsed, inductive current drive [Luc11]. However, latter current is a mandatory requirement for efficient steady-state operation, since the solitary usage of neutral beam injection and ECCD is highly inefficient. Until then, the tokamak remains a pulsed reactor, depending on additional poloidal field coils. The number of required toroidal and poloidal field coils depends on the volume and the aspect ratio (a:R_0) of the desired plasma.

An exemplary magnet structure of a tokamak with its poloidal, toroidal and solenoid magnet coils is visualized in the following figure 2.8.

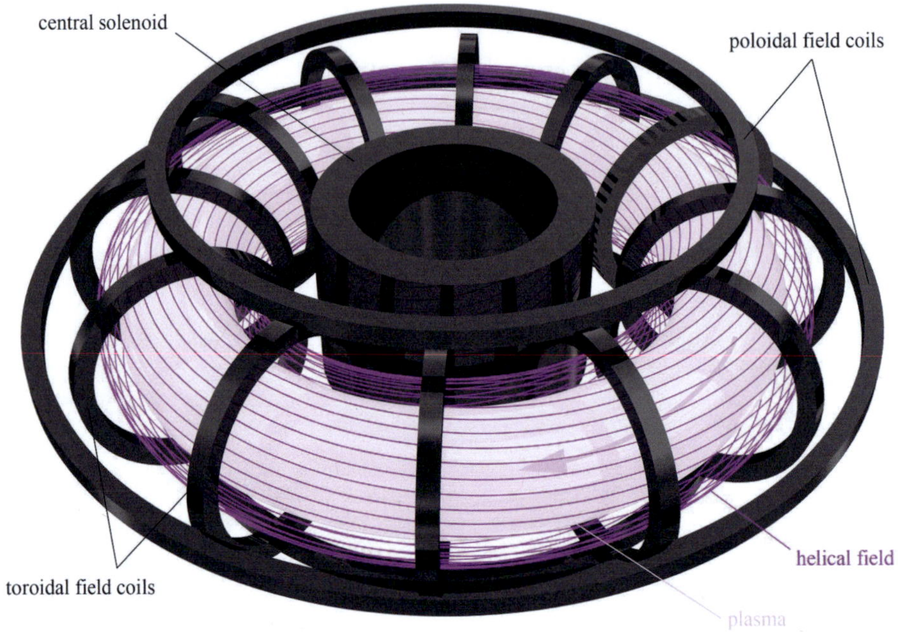

Figure 2.8: Magnetic plasma confinement of a tokamak. The helically wound, magnetic torus yields from superposition of toroidal and poloidal magnetic fields.

Although the tokamak design is not operated in steady-state mode, the tokamak is qualified to be the first fusion power station candidate for following reasons:

- Main advantage of the tokamak over the stellarator is its high plasma stability. While in a stellarator fast particles tend to leave the asymmetric, magnetic confinement and thereby reduce the plasma temperature, in a tokamak, with a rotationally symmetric magnet torus, this effect plays a minor role.
- As the toroidal field in a tokamak is induced by large-scale planar coils with a rotationally symmetric reactor chamber, the construction and maintenance of the machine is less sophisticated, since all components are far better accessible. Especially the frequent exchange of the breeding blankets in a fusion power station requires a straight forward, brief workflow to keep the shutdown period to a minimum.

2.3.1 International Thermonuclear Experimental Reactor (ITER)

With a fusion power of 500 MW and a plasma volume of 837 m^3, the International Thermonuclear Experimental Reactor, ITER, will be the largest tokamak ever built. The first development agreement was reached in 1986 between the European Union, Japan, the Soviet Union and the USA. Ever since, 35 countries joined the ITER agreement, which was officially signed in November 2006. One year later, the ITER organisation was officially established. First on-site construction already began in 2005 at Cadarache, France, with the goal of first plasma ignition expected in 2025[1].

Due to the required major plasma radius of $R_0 = 6.2\,\mathrm{m}$ and the minor plasma radius of $a = 2.0\,\mathrm{m}$, the magnetic confinement has to be generated by superconducting large-scale magnet coils. While the toroidal field is induced by 18 D-shaped Nb_3Sn coils, the poloidal field is induced by 6 stacked Nb_3Sn solenoids and 6 circular NbTi coils. Eighteen copper correction coils outside the vacuum vessel allow for additional plasma stabilisation. An additional array of 27 in-vessel NbTi coils around the reactor chamber allows for stabilisation against edge-localized mode (ELM) instabilities, which can severely damage the reactor wall. A visualisation of the ITER magnet system is shown in figure 2.9. Summarized parameters on the reactor are found in table 2.1.

With its fusion power of 500 MW and heating power of 50 MW, ITER will be the first reactor experiment expected to reach a positive return on energy ($Q \geq 10$) [ABS02]. ITER is supposed to reach the point of plasma ignition and maintain the self-sustaining plasma for various minutes. Although ITER is no self-breeding reactor, it is the key facility to investigate tritium breeding, neutral beam injection, wave induced current drive, large-scale tritium handling and remote handling for maintenance of fusion reactors. All experience and knowledge gained from ITER is essential to lead the way to a first demonstration fusion power plant.

1 According to the ITER Resource-Loaded Integrated Schedule, released in June 2016 [Cre].

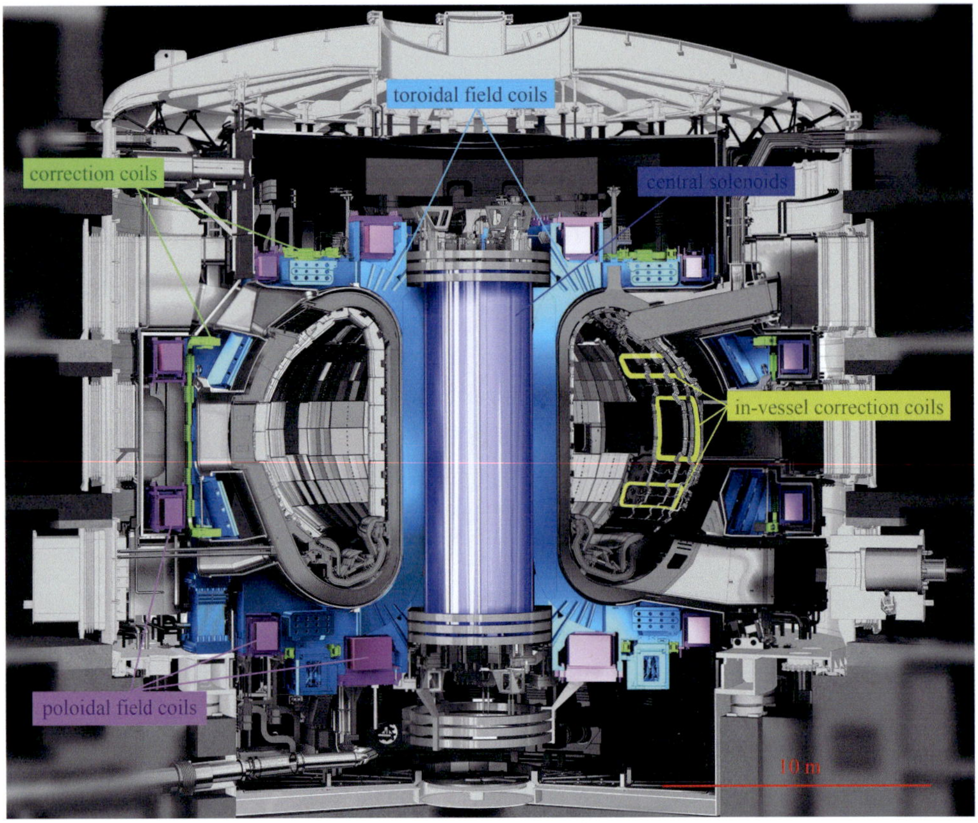

Figure 2.9: Magnet components of ITER. The entire system consists of 18 toroidal, 6 poloidal, 18 correction and 27 in-vessel correction coils and 6 stacked central solenoids. Except for the Nb_3Sn toroidal and central magnets, all coils are made from NbTi. Original picture from [Cre].

2.3.2 European DEMOnstration Power Plant (DEMO)

The DEMOnstration Power Plant, DEMO, is seen as the first approach to an industrial and economical tokamak fusion reactor to qualify nuclear fusion as feasible alternative to any hitherto fossil power source. With 35 countries participating in the ITER project and some of these countries designing their own DEMO reactor, the range of different DEMO designs and concepts is wide spread. However, as this thesis is focusing on the work of the *European Consortium for the Development of Fusion Energy* and the *European Joint Undertaking for ITER and the Development of Fusion Energy*, in all following chapters, the term DEMO refers to the calculated European DEMOnstration Power Plant design called DEMO1.

According to the official EUROfusion road map [Rom12], the construction and first operation of DEMO is planned for 2030 to 2040, with demonstration of fusion electricity by 2050. Such goal means

- to reach sufficient plasma stability to operate a DEMO in pulsed or steady-state mode,
- to reach tritium self-sufficiency, which implies autonomous tritium breeding inside the blankets,
- to demonstrate a positive and economical return on energy,
- and to resolve all technological challenges given by the extreme operation conditions

within the next 34 years.

DEMO aims for a fusion power of 2000 MW with a net electrical power of about 500 MW, while ITER has only quarter the fusion power and a negative net electrical power. A detailed comparison of the key features of both facilities is given in table 2.1.

In order to be able to satisfy the technical requirements, it is mandatory to look for new technologies and better solutions as they have been used for ITER. Naturally, this includes the entire magnet system, which is required to confine the plasma with a volume of two to three times of the ITER plasma. Such enormous dimensions require larger toroidal field coils with higher current densities or a larger number of coil windings to induce the required magnetic field of up to 6.85 T at the centre line of the plasma torus. Either of both upgrades implies thicker and more voluminous toroidal field coils with higher peak magnetic fields at the conductor. Although these requirements could be met with conventional Nb_3Sn low temperature superconductors, technical high temperature superconductors enable higher currents densities, higher peak magnetic fields, higher operation temperatures, an increased temperature safety margin, higher tensile strength and pressure durability and an increased flux-jump stability.

These advantages of high temperature superconductors motivate the investigation of various high temperature superconductor cable designs for application in future DEMO toroidal field coils, beginning with following introduction to high temperature superconductors and an overview of hitherto large scale magnet technology.

Table 2.1: Main parameters of Wendelstein 7-X, ITER and DEMO. Data from [ABS02, WBDW16, Kem12, Kem15, Cre, Dol13].

parameters	W 7-X	ITER	DEMO
fusion reaction	$^2H + ^2H$	$^2H + ^3H$	$^2H + ^3H$
total fusion power	-	500 MW	2000 MW
gross electrical power	-	< 160 MW	900 MW
net electrical power	-	< 0 MW	500 MW
heating / current drive power	10 MW	50 MW	50 MW
fusion gain factor Q	-	10	40
plasma β-value	4-5%	1.8% (normalised β)	2 - 2.6% [2] (normalised β)
plasma volume	30 m³	837 m³	1450 - 2500 m³ [2]
major plasma radius R_0	5.5 m	6.2 m	9 m
minor plasma radius a	0.53 m	2.0 m	2.25 - 2.9 m [2]
aspect ratio	10.4	3.1	3.1 - 4 [2]
number of modules	5	-	-
number of superconducting toroidal field coils	50	18	16
number of superconducting poloidal field coils	-	6	6
number of superconducting central solenoids	-	6	unknown
number of copper correction coils	20	18	unknown
number of superconducting in-vessel coils	-	27	unknown
operating current in toroidal field coils	17.6 kA	67.91 kA	50 kA
magnetic field at plasma	2.5 T	5.3 T	5.67 - 6.85 T [2]
entire facility costs	10^9 EUR	13×10^9 EUR	probably below ITER

[2] Depending on the DEMO1 July 2012 or April 2015 design.

3 Fusion magnet technology

3.1 Superconductors for fusion technology

Superconductivity describes the ability of elements, molecules or chemical compounds to overcome electric resistivity when cooled below a characteristic critical temperature T_c. This phenomenon was discovered in April 1911 by Heike Kamerlingh Onnes when he measured the electrical resistivity of mercury, tin and lead [Onn11]. The change in various physical properties, when switching from normal to superconducting state, implies a phase transition of second order [LAN36].

3.1.1 Critical magnetic field

In technical high temperature superconductor applications the Meißner phase of the superconductor plays a minor role. The same accounts for the critical magnetic field B_{c1} and B_{c2}. Instead, the focus lies on the irreversibility field B_{irr}, which is the maximum magnetic field at which the bulk current density of the superconductor becomes zero [LGFP01]. The irreversibility field is often simply referred to as B_c. According to [Kom95], it can be approximated by

$$B_{\mathrm{irr}}(T) = B_{\mathrm{irr}}(0\,\mathrm{K}) \cdot \left(1 - \left(\frac{T}{T_c}\right)^2\right)$$

with the sample temperature T and the absolute critical temperature T_c.

Experience has shown, that for some technical high temperature superconductors the better approximation is

$$B_{\mathrm{irr}}(T) = B_{\mathrm{irr}}(0\,\mathrm{K}) \cdot \left(1 - \frac{T}{T_c}\right)^{\alpha}$$

with α as arbitrary fit parameter [Wes10, UBS+15].

The magnetic field acting on the superconductor is either self-induced or externally induced. A self-induced field occurs in any superconducting strand and cable compositions and is referred to as *self-field*. Except for coil windings, the self-field is normally of few millitesla, which is negligible if operated inside a typical magnetic background field of few tesla. The externally induced field also known as *background field* comes from external steady state electromagnets which are able to generate up to 45 T as hybrid magnets and 27 T as superconducting magnets [Mil03, RW15]. With pulsed electromagnetic flux compression it is even possible to achieve magnetic fields of several hundred tesla [SMI+04]. In this thesis, the experimental peak magnetic field is up to 12 T and is induced by a superconducting split-coil magnet. Technical details are described in chapter 5.

3.1.2 Critical current density

The critical current density j_c is defined as the maximum current density before a superconductor becomes normal conducting. It is affected by the irreversibility field B_{irr} and the temperature T of the superconductor. The schematic function $j(B,T)$ for a rare-earth-barium-copper-oxide (*RE*BCO) superconductor is plotted in figure 3.1.

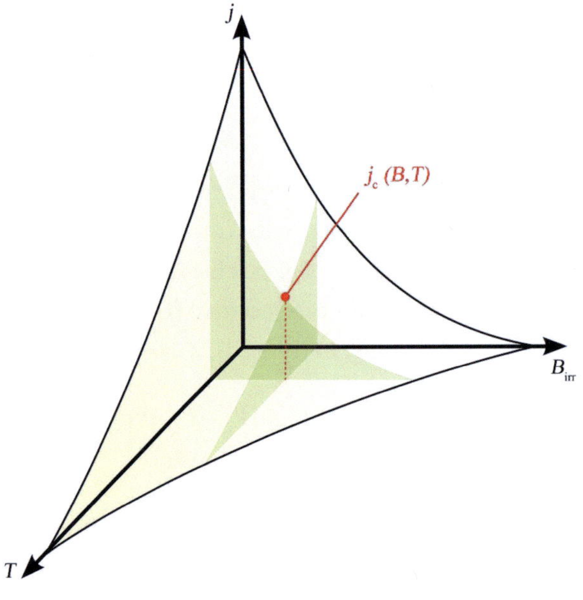

Figure 3.1: Schematic function $j_c(B,T)$ depending on the magnetic irreversibility field B_{irr} and the conductor temperature T for a rare-earth-barium-copper-oxide (*RE*BCO) superconductor.

Here, the critical current density $j_c(B,T)$ is represented by the surface spread between the three axis. With increasing field or conductor temperature, the current density drops significantly.

For first generation (1G) high temperature superconductors (HTS) bismuth-strontium-calcium-copper-oxide BSCCO-2212 and BSCCO-2223 the critical current density is defined as the current density which induces an electric field along the superconductor of 1 μV per 1 cm conductor length [DKE06]. A standardisation for second generation (2G) high temperature superconductors (HTS) respectively rare-earth-barium-copper-oxide *RE*BCO tapes, is not yet established [VDE16] . Therefore, the $E_c = 1\,\mu\text{V/cm}$-criteria was adapted to all measurements with 2G superconductors within this thesis.

A mathematically convenient approximation to describe the superconducting transition for technical superconductors is the empiric power function

$$E = E_c \left(\frac{j}{j_c} \right)^n \tag{3.1}$$

called *power law*, which depends on the measured electrical field E, the current density j [BK13, Kom95], the critical electric field E_c and the critical current density j_c. Although it does not perfectly describe the complete transition to normal state, it precisely approximates the transition close to the critical electric field E_c. Figure 3.2 schematically demonstrates the E-j dependency.

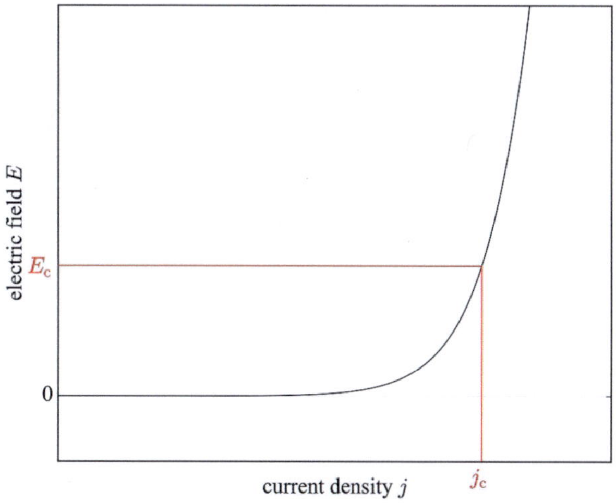

Figure 3.2: Schematic of a superconducting transition in a technical type II superconductor. The measured transition can be approximated for values close to the critical electric field E_c by the power law function.

It is important to follow a standard method to estimate the critical current density j_c from superconductor measurements. A standard procedure was established and realized in a python script. In the first step, the measured data is corrected by a possible linear ohmic contribution. This is recommended for measurements where the voltage taps are not directly connected to the superconductor tapes, but to the copper casing. The ohmic contribution of the copper between the superconductor tapes and the voltage taps needs to be subtracted to estimate the critical current precisely.

The second step is the calculation of the critical current density j_c. Since a double logarithmic scaling with any base of a monomial power function results in a linear function, both, the current density and the electric field can be visualised as demonstrated in figure 3.3.

In order to estimate the critical current I_c precisely, the linear function is fitted by a polynomial least square fit of first order

$$ln(E) = n \cdot ln(j) + c \tag{3.2}$$

with the fit parameters n and c.

Since the power law only accounts for values close to the critical electric field E_c, it is convenient to consider the values shortly before and those after the 1 μV/cm criteria. By solving function 3.2 with $E = E_c = 1\,\mu\text{V/cm}$, the critical current density j_c respectively the critical current I_c can be calculated by

$$j_c = e^{\frac{ln(E_c)-c}{n}} = e^{-\frac{c}{n}}.$$

Although it is also possible to simply read the critical current from a graph, this method provides a much more precise result - especially for superconducting transitions in magnetic background fields or sample temperatures above 4.2 K.

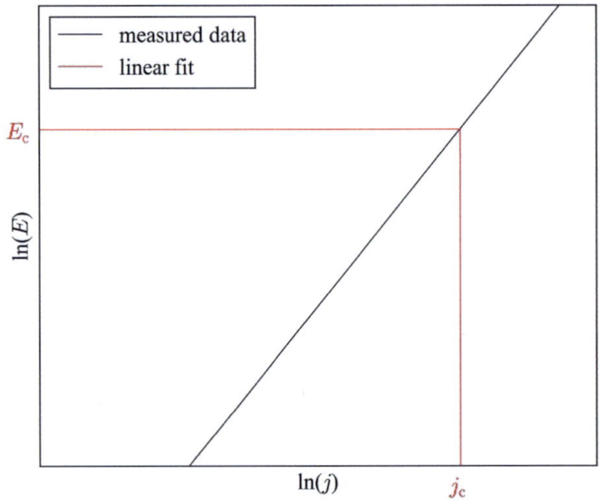

Figure 3.3: Schematic of a superconducting transition in a technical type II superconductor. Both, electric field E and the current density j are scaled logarithmically. The expected polynomial fit around the critical electric field E_c is of first order.

3.1.3 *n*-value

As seen in equation 3.2, the steepness of the power law function correlates with the n-value. Typical n-values for 2G *RE*BCO tapes are between 30 and 40 for self-field at 77 K, depending on the manufacturer and the wire specifications. Values below often indicate a flat superconducting transition due to external magnetic fields, temperature variations, or - in the worst case - sample degradation [Bru04]. Therefore, the n-value is a required indicator to evaluate the quality of a superconducting transition.

The estimation of the n-value within this thesis is executed in the same way as the estimation of the critical current. The E-j-function is scaled double logarithmically to base e and fit by equation 3.2. The n-value fit considers all values between $E_c = 1\mu V/cm$ and $E_c = 10\mu V/cm$.

3.1.4 *Rare-Earth*-Barium-Copper-Oxide (*RE*BCO) superconductors

Rare-Earth-Barium-Copper-Oxide (*RE*BCO) superconductors are type II superconductors. Due to their high critical temperature $T_c > 77\,K$ they belong to the high temperature superconductors. The first discovered *RE*BCO superconductor was a Y-Ba-Cu-O composition, published in 1987 by Wu et al. [WAT$^+$87]. With further development, yttrium was substituted by other rare earth elements such as gadolinium, samarium, disprosium or lanthanum. Nowadays, most commercially available *RE*BCO superconductors either use yttrium, gadolinium, samarium or disprosium as core elements.

*RE*BCO superconductors are cuprate compounds where only the CuO_2 monolayers are superconducting. All other layers are isolation layers. A schematic view of the orthorhombic crystal system with its three axes is shown in figure 3.4 [MW06].

The maximum critical temperature of *RE*BCO superconductors depends on the oxygen concentration. The highest temperature values around $T_c \approx 91\,K$ are achieved for $REBa_2Cu_3O_{7-\delta}$ with an oxygen content of $0 \leq \delta \leq 0.2$ [MW06]. For oxygen contents of $0.5 \leq \delta \leq 1$, *RE*BCO superconductors become antiferromagnetic insulators [MW06]. The oxygen concentration can be increase or reduced by annealing the *RE*BCO superconductor inside oxygen atmosphere.

Typical coherence lengths for type II *RE*BCO superconductors in ab-direction are $\xi_{ab}(0\,\mathrm{K}) = 2.5$ nm and in the c-direction $\xi_c(0K) = 0.5$ nm. The isolating yttrium, barium-oxide and copper-oxide layers are responsible for the low coherence length in vertical direction.

Commercially available *RE*BCO

Superconducting *RE*BCO tapes, also known as coated conductors, are meanwhile commercially available, offered by several manufacturers from different countries all over the world. Independent of the manufacturer, all coated conductors have a flat tape structure, as demonstrated in figure 3.5, in common.

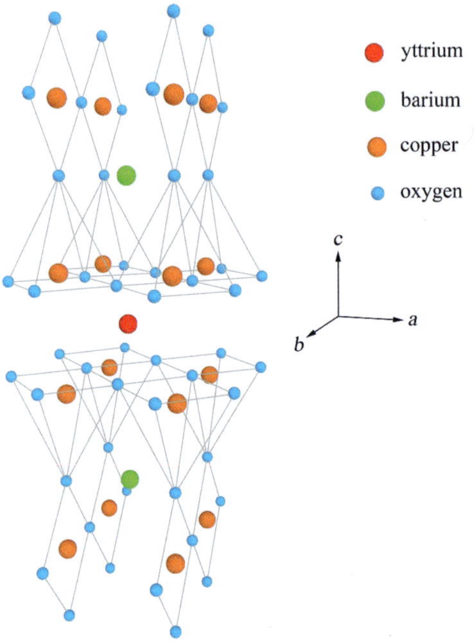

Figure 3.4: Crystal lattice of YBCO.

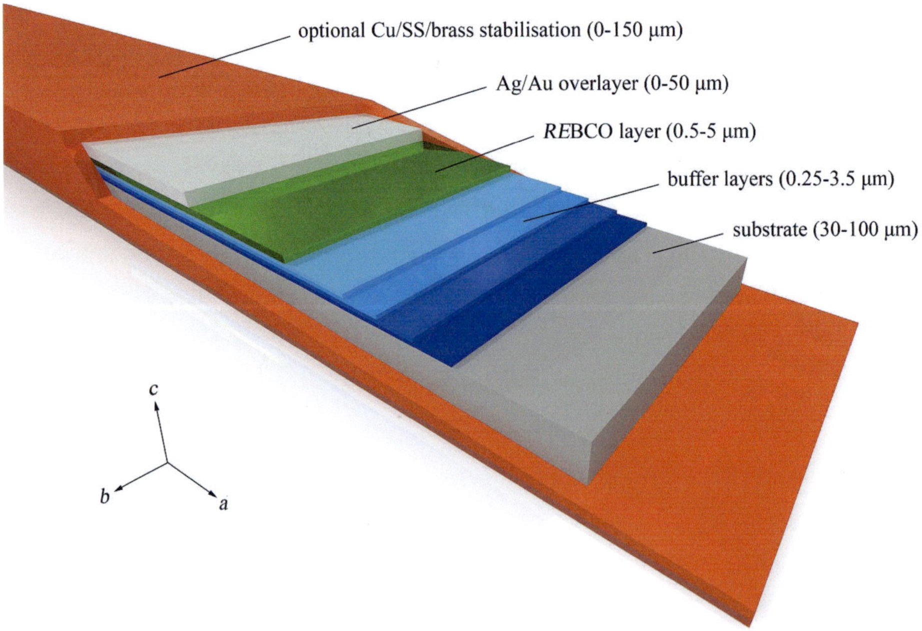

Figure 3.5: Tape structure of a *RE*BCO coated conductor.

The base material of the *RE*BCO superconductor is the substrate. It is made of Hastelloy©C276 or stainless steel with a thickness between 50 and 100 micrometers. On top of the substrate, several buffer layers are deposited to act as mechanical and chemical diffusion barrier between the ultimate *RE*BCO layer and the substrate. These buffer layers consist of metal-oxides or ceramics such as magnesium oxide (MgO), yttria (Y_2O_3) and yttria-stabilized zirconia (YSZ). Typical buffer layer thicknesses vary between 0.25 and 3.5 micrometers. Since the crystal lattice of the substrate is different from the lattice of the superconducting *RE*BCO layer, either the substrate or the buffer layers need to be textured to allow for an aligned and homogeneous *RE*BCO growth. Rolling assisted biaxially textured substrates (RABiTS) already come with the required, textured surface. Alternatively, buffer layers can be textured by alternating beam-assisted deposition (ABAD), ion beam-assisted deposition (IBAD) or inclined substrate deposition (ISD).

The superconducting *RE*BCO layer is grown on top of the buffer layers with its c-axis perpendicular to the tape plane. Slight angle variations between the c-axis and the perpendicular to the tape plane can be observed for tapes from different manufacturers. With a thickness of only one to three micrometers, the *RE*BCO layer is very thin, but reaches current densities of up to 100 $\mathrm{kA/mm^{-2}}$ at 4.2 K in a 19 T magnet background field [SBB$^+$16]. Commercial *RE*BCO layers are grown by different methods: metal organic chemical vapour deposition (MOCVD), metal organic decomposition (MOD), pulsed laser deposition (PLD), physical vacuum deposition (PVD), chemical solution deposition (CSD), reactive co-evaporation and cyclic deposition reaction (RCE-CDR) or the faster reactive co-evaporation and deposition reaction (RCE-DR) [Moo14, IOF$^+$93, Sup, SRM$^+$08, Ame15, NZB$^+$15].

In a final step, the *RE*BCO layer is coated with a silver or gold overlay. The thickness of this overlay highly depends on the manufacturer and varies between zero and 50 micrometers. It stabilises the superconductor thermally and electrically and protects the *RE*BCO layer against mechanical damage. To further improve the electrical stabilisation or the applicability of the coated conductors, most manufacturers offer an additional copper, brass or stainless-steel stabilisation which is electroplated, laminated or soldered on or around the superconductor tape. To further increase the applicability of the coated conductors, some manufacturers even offer to pre-tin the superconductor tapes with soft-solder or to laminate them with isolating polyimide. The specifications of different, commercial coated conductor tapes are summarized in table 3.1.

Mechanical properties

Due to their Hastelloy© or stainless steel substrates, coated conductors have good mechanical properties. They show a low strain dependency. In-field measurements of the critical current as a function of longitudinally applied tensile strain at 4.2 K inside a 12 T background field are shown in figure 3.6. The maximum tensile strain before irreversible degradation is listed, for each coated conductor individually, in table 3.1.

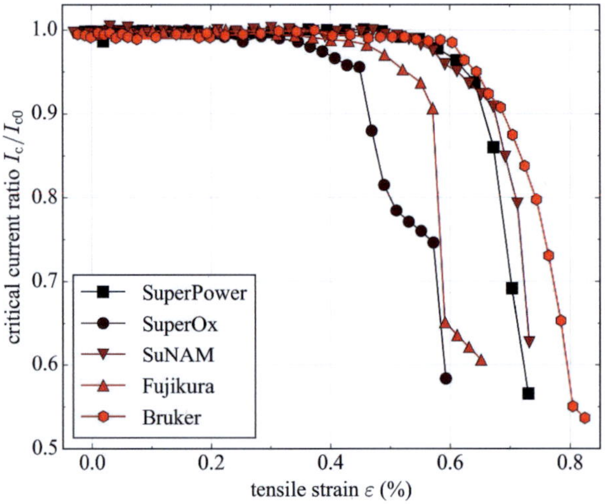

Figure 3.6: Current-tensile-strain function for various commercial REBCO superconductor tapes at 4.2 K inside a 19 T magnet background field. Data from [BMS15].

All investigated superconductors withstood tensile strain up to 0.4% under assumption of a 5% critical current tolerance. Coated conductors provided by SuperPower, Bruker and SuNAM even reached a strain level of 0.6%. Although all tapes demonstrate a solid strain resistance, after all, the absolute strain level is still very dependant on the manufacturer. This dependence also accounts for the stress-strain function of coated conductors. The tensile strain-stress functions at 4.2 K are plotted in figure 3.7. The corresponding Young's moduli and yield strengths are listed in table 3.1.

Table 3.1: Specifications of commercially available *RE*BCO coated conductors. Data from: SuperPower [NZB+15, Bar15, BMS15], SuperOx [Ame15, Cel, Bar15, BMS15], SuNAM [Moo14, Bar15, BMS15], AMSC [SRM+08, Bar15], Theva [THE14a, THE14c, THE14b], Bruker [Bru13, Bar15, BMS15], STI [Sup], Fujikura [Iij13, YYO11, Bar15, BMS15], d-nano [Deu16].

properties	SuperPower	SuperOx	SuNAM	AMSC	Fujikura	Theva	STI	Bruker	d-nano
place of manufacture	USA	Russia	South Korea	USA	Japan	Germany	USA	Germany	Germany
manufacturing process	MOCVD	PLD, MOD	RCE-DR	MOD	PLD	RCE-CDR	RCE-CDR	PLD	PVD, CSD
texturing	IBAD	IBAD	IBAD	RABiTS	IBAD	ISD	IBAD	ABAD	-
substrate material	Hastelloy© C276	Hastelloy© C276	Hastelloy© C276, SS	Hastelloy© C276	Hastelloy© C276	Hastelloy© C276	Hastelloy© C276, SS	Hastelloy© C276, Cr-Ni SS	Cube textured Ni5W, Ni9W
substrate thickness (μm)	30, 50, 100	60, 100	60, 80	50-75	75, 100	60, 100	100	50, 100	60
buffer layer thickness (μm)	1	< 0.4	1	0.25	1.18	1, 3.5	3	1.55	0.01-0.15
superconductor material	YBCO	GdBCO	GdBCO	YBCO	YBCO, GdBCO	GdBCO	YBCO	YBCO	YBCO
superconductor thickness (μm)	1	1-3	1-3	1	2	1.5, 2	3-5	0.5-3	0.5-1
silver overlayer thickness (μm)	1-5	0-10	1.5	1	20	3	1-50	0.2 Ag, Au	-
stabilisation material	Cu	Cu, PbBiSn, PbSn, SnIn	Ag, Cu, brass, SS	Cu, brass, SS	Cu	Cu	Cu, Ag, brass	Cu	Cu
stabilisation thickness (μm)	10-115	0-100	60	75-150	75, 100	20	0-150	0.1-40	5-20
width (mm)	2, 3, 4, 6, 12	4, 6, 12	2-10, 12	4, 5, 12	5, 10, 12	12	3, 4, 10	4, 12	4, 10, 12, 20
max. I_c(77 K) (A/cm-width)	420	420	620	420	500	300	500	272.5	250
j_{eng}(4.2 K, 12 T ⊥) (A/mm^2)	751	552	278	102	389	-	-	751	-
j_{eng}(4.2 K, 13 T ⊥) (A/mm^2)	707	525	265	96	370	-	327 (15 T)	700	-
max. single piece length (m)	500	500	300	440	500	50	-	300	100
Young's modulus at 4.2 K (GPa)	155	172	168	-	176	-	-	187	-
yield strength at 4.2 K (MPa)	893	1083	846	-	787	-	-	736	150, 250
ε_{irr}(4.2 K, 12 T ⊥) (%)	0.66-0.68	0.45-0.47	0.67-0.69	-	0.55-0.57	-	-	0.7-0.72	-
σ_{irr}(4.2 K, 12 T ⊥) (MPa)	800-830	770-800	830-840	-	750-760	-	-	740-750	-

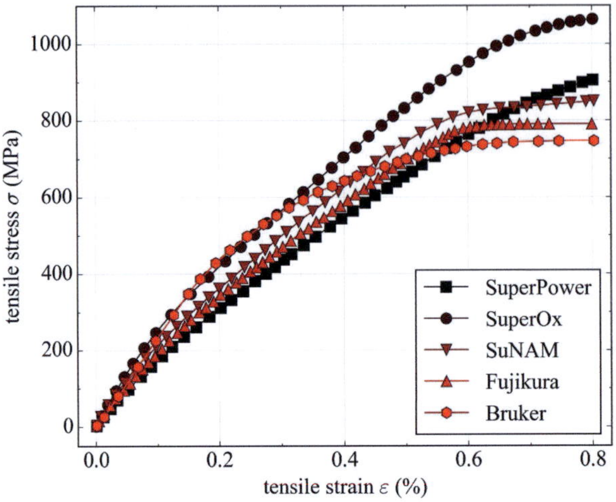

Figure 3.7: Tensile-stress-strain function for various commercial *RE*BCO superconductor tapes at 4.2 K. Data from [BMS15].

With a Young's modulus of 155 GPa and a yield strength of 1083 MPa, SuperOx tapes have a steeper stress-strain curve than any other coated conductor. Bruker tapes have with 736 MPa the lowest yield strength. SuperPower tapes show the highest elasticity with a Young's modulus of only 155 GPa and for stress below 700 MPa a higher plasticity than any other coated conductor. The different stress-strain behaviour of commercial coated conductors directly influences their resulting current-stress behaviour which is plotted in figure 3.8.

Although all coated conductors showed different current-tensile-strain curves, their individual plasticity leads to flat and very similar current-tensile-stress functions. Within the small spread of 740 MPa to 840 MPa, all coated conductors degrade irreversibly. For tensile stress below 740 MPa, all tested commercially available coated conductors still carry more than 95% of the original current and for stress below 600 MPa, the critical current ratio is above 99%. The maximum tensile stress before irreversible degradation is listed, for each coated conductor individually, in foregoing table 3.1.

In addition to their very high durability against tensile stress, commercially produced coated conductors also come with a high durability against hydrostatic pressure. Peak pressure values of more than 200 MPa do not impair the current density of *RE*BCO coated conductors [FQD+12]. The only risk of destroying the *RE*BCO coated conductors severely are delamination effects, where the *RE*BCO layer is torn off the substrate. However, as long as the *RE*BCO tapes are embedded inside a conduit and therefore protected from arbitrary movement and bending, delamination is not an issue.

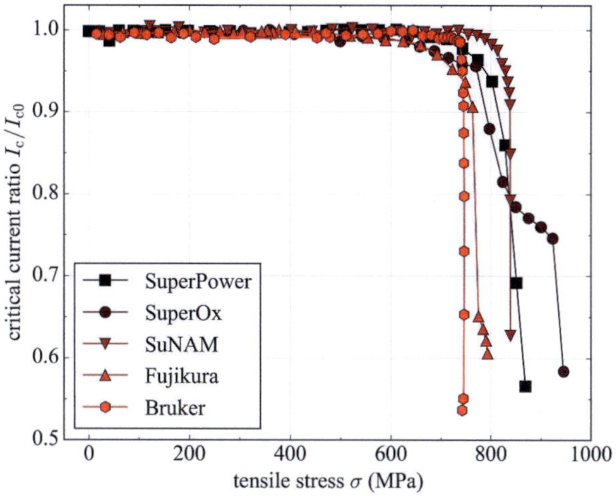

Figure 3.8: Current-tensile-strain function for various commercial *RE*BCO superconductor tapes at 4.2 K inside a 19 T magnet background field. Data from [BMS15].

Magnetic field dependence

In-field critical current measurements of coated conductors highly depend on the angle α between the c-axis of the crystal lattice and the magnet flux. Usually, lowest current densities are measured for magnetic field lines parallel to the c-axis. With increase of the angle, the current density rises. Peak values are achieved for angles around $\alpha = 90°$. The precise angular dependency of 2G superconductors is different for each manufacturer and changes with new developments on the doping process of artificial pinning centres during the *RE*BCO growth.

However, inside a DEMO TF magnet coil, the angular dependency of the critical current density inside coated conductors plays a minor role. Since it is impossible to arrange all coated conductors of a twisted and transposed full-size TF coil cable, it is impossible to avoid coated conductors being arranged perpendicularly to the magnetic field lines. Therefore, all in-field measurements in this thesis were executed with the superconducting samples preferably arranged with the c-axis parallel to the magnetic field. Up-scaling calculations for full-size cables are also executed on this basis.

3.2 ITER TF coil design parameters

As this thesis focuses on high temperature superconductor cable concepts for the toroidal field coils of tokamak reactors, the discussion on toroidal field magnet technology is limited to the present tokamak experiment ITER and the future tokamak demonstration reactor DEMO. As ITER is already under construction, its low temperature superconductor state-of-the-art magnet technology is summarized and will be used as reference for calculations on the DEMO high temperature superconductor toroidal field coil design.

The ITER magnet coil architecture consists of 18 circular arranged toroidal field coils with a total, stored energy of 41 GJ [ITE06]. The entire current per coil integrated over the number of windings is 9.1 MA, which induces the required magnetic field of 5.3 T along the 34.1 m long central line of the plasma torus [ITE06, Sip04]. For technical reasons, each toroidal field coil has a D-shaped form. This form has the lowest in-plane stress and provides sufficient space for the central solenoid, while satisfying the defined plasma aspect ratio, which is the ratio between the minor plasma cross-section radius a and the major radius of the entire plasma torus R_0. For ITER the expected values are $a = 2.0\,\mathrm{m}$ and $R_0 = 6.2\,\mathrm{m}$, which corresponds to an aspect ratio of 3.1 [Sip04]. The dimension of an ITER toroidal field coil, to fulfil the magnetic field parameters demanded from plasma physics side, is demonstrated in figure 3.9. With a diameter of 12.6 m and a width of 8.1 m, the coil winding is very large compared to previous fusion devices. The D-shaped coil consists of the vertical inboard leg and the curved outboard leg. With a distance of less than three meters, the inboard leg is very close to the centre axis of the torus. Due to their isosceles, trapezoidal cross-section, the inbound legs of all 18 toroidal magnet coils form together the centre column of the reactor chamber. The outboard leg is not trapezoidal, but rectangular and serves additionally as support structure for four of the six poloidal field coils.

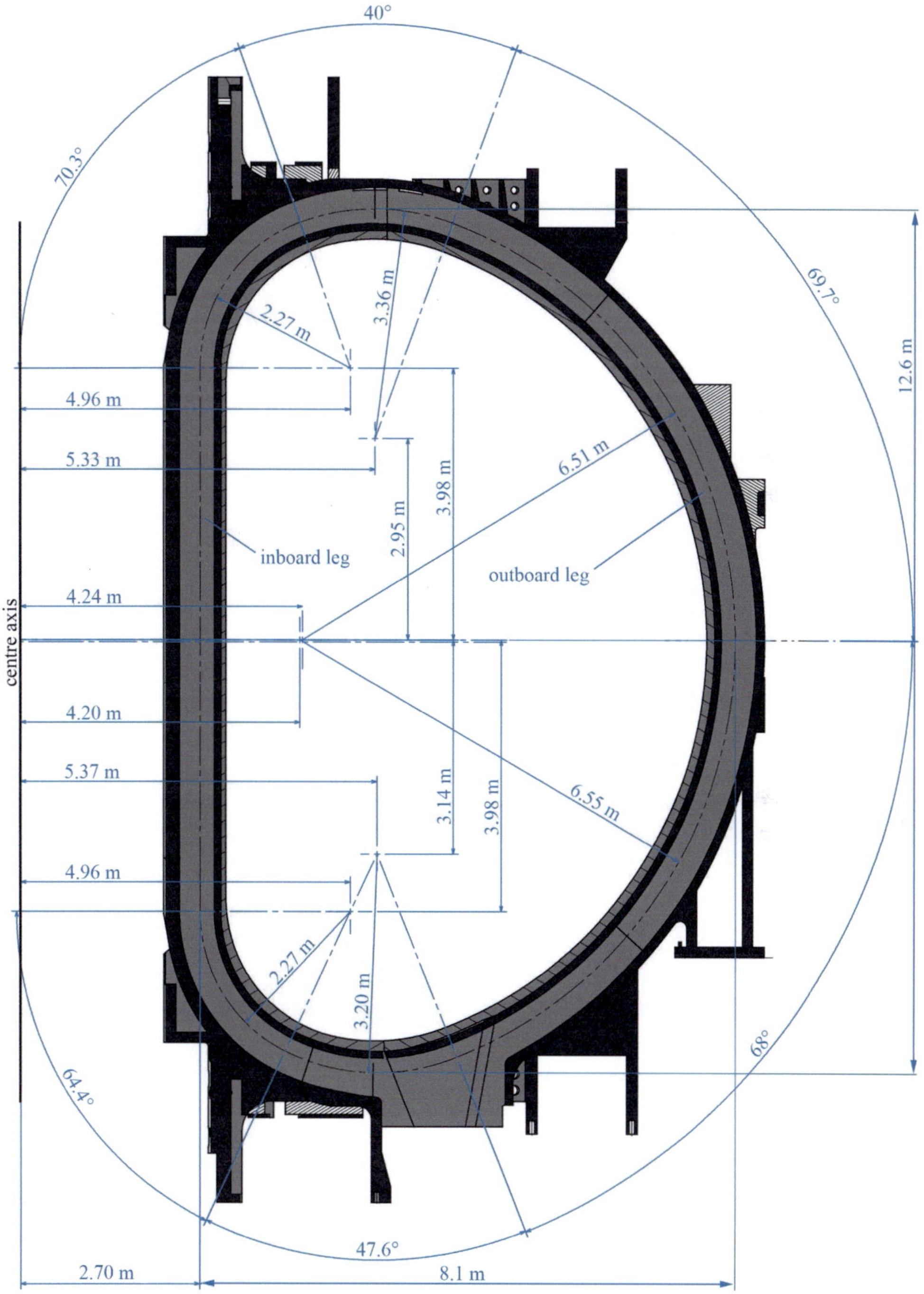

Figure 3.9: Dimensions of an ITER toroidal field coil. Picture after [ITE06].

3.2.1 Winding pack

The winding pack contains the wound conductors and is embedded inside the inboard and outboard leg, surrounded by a massive austenitic stainless steel structure. A cross-section view is shown in figure 3.10. The winding pack consists of seven radial plates, each with a double pancake winding, which results in a total number of 14 pancakes. The maximum number of turns per pancakes is 11. The outermost pancakes only have three and nine turns respectively, to adept to the trapezoidal cross-section of the inboard leg. In sum, the winding pack consists of 134 conductor turns. With the entire coil current of 9.1 MA, each conductor winding must carry 67.91 kA. The peak magnetic field is estimated to be around 11 T at the innermost coil windings, experienced by the conductor as a gradient from 10.3 T to 11.9 T, which is equivalent to a homogeneous background field of 10.8 T [BM08, BBC+08].

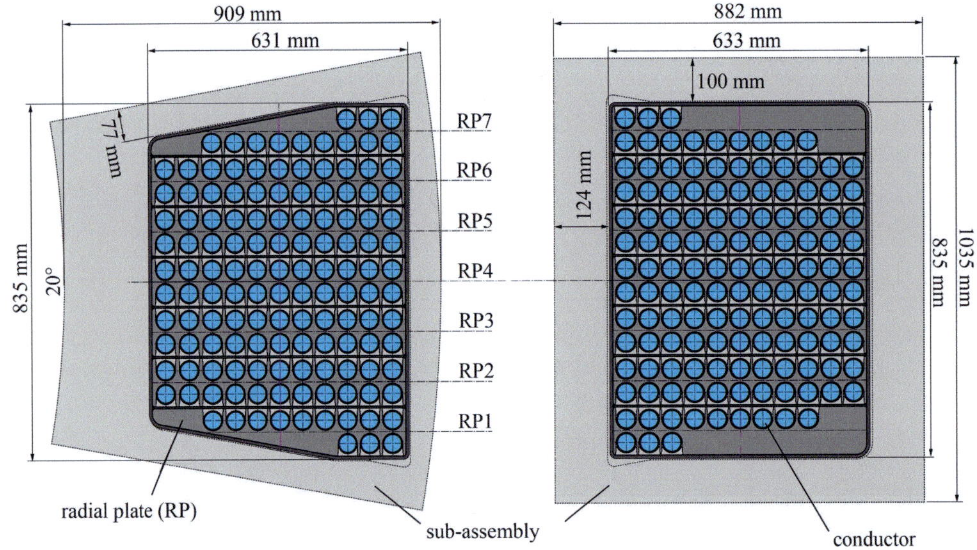

Figure 3.10: ITER winding pack for inboard (left) and outboard leg (right). Picture after [ITE06].

3.2.2 Cable in Conduit Conductor (CICC)

In order to satisfy the operation current requirement of 68 kA inside a conductor without ohmic losses, the conductor is manufactured from superconducting Nb_3Sn. It consists of 900 chromium coated superconductor- and 522 copper strands. To reduce AC losses, all strands have a diameter of only 0.82 mm and are twisted several times during the manufacturing process. The exact number of twists and the corresponding twist pitch is demonstrated in figure 3.11.

In step five, five bunches with 30 Nb_3Sn and 15 copper strands each, are twisted around a

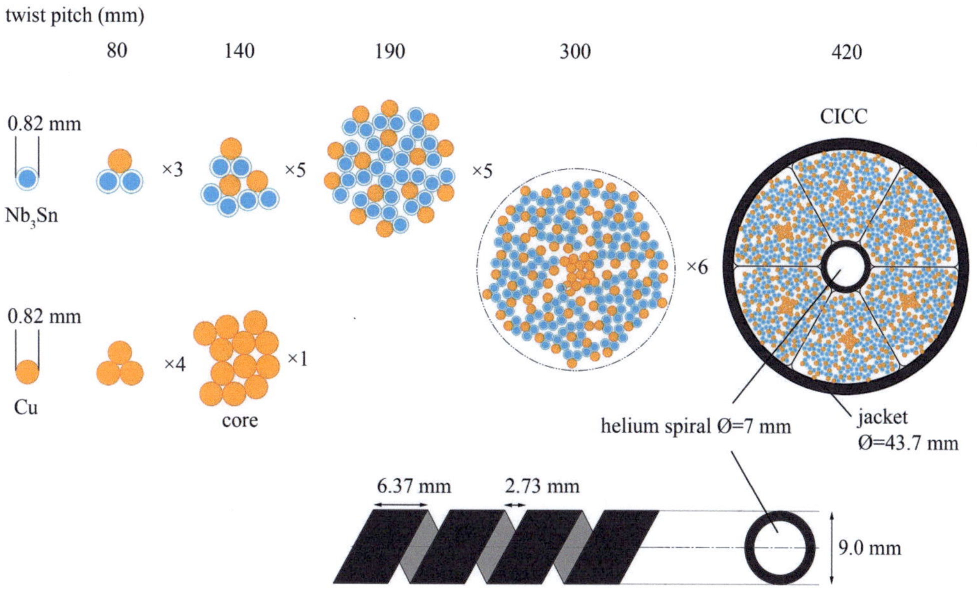

Figure 3.11: ITER Cable in Conduit Conductor (CICC). Twist pitch lengths from [BM08]. Picture after [ITE06].

core of 12 twisted copper strands. Six of the resulting bunches are wrapped with 0.05 mm stainless steel foil and afterwards twisted around a stainless steel spiral with an outer diameter of 9 mm, which provides additional helium void to reduce the required force flow pressure. Before the cable is pulled and compacted inside its steel jacket, it is wrapped once more into 0.08 mm thick stainless steel foil. The outer diameter of the Cable in Conduit Conductor (CICC) for the toroidal field coil is 43.7 mm and the overall CICC length per winding pack is 4570 m [MDL$^+$12].

For later comparison between ITER and DEMO toroidal field coil conductors, the material ratio of superconductor, electrical copper stabilisation and mechanical reinforcement is of high importance. The cross-section areas of all components and their corresponding fraction of the entire winding pack area inside the inboard leg are listed in table 3.2.

The CICC conductor is cooled by forced flow cooling. Supercritical helium circulates inside a flow loop between a heat exchanger and the coil structure. In order to penetrate the narrow CICC, it is pressed through the central helium spiral and the helium voids of the CICC with an inlet pressure of 0.6 MPa [ITE13], that drops by 0.08 - 0.15 MPa along the coil windings. The flow rates are about 2 - 3 kg/s and the helium temperature varies between 4.35 K and 4.5 K [ITE13]. The maximum CICC temperature at the peak magnetic field of 11.8 T is expected to be 5.0 K [ITE06]. In case of unforeseen heat input, it is possible to further reduce the operation temperature of the heat exchanger from 4.2 K to 3.7 K [ITE06].

Table 3.2: Cross-section area of the inboard leg sorted by material component. Data from [ITE09a, ITE09b].

cross-section area of	entire inboard leg (mm^2)	per coil winding (mm^2)	ratio (%)
Nb_3Sn	63650	475	8.2
copper stabilisation	70216	524	9.0
helium void	43148	322	5.6
stainless steel (only jacket)	35108	262	4.5
stainless steel (plates, jacket, casing)	562000	4194	72.4
insulation	> 37000	> 276	4.8

3.2.3 Safety factor

Beside the typical operational heat input due to ohmic heat at the cable joints, thermal conductivity of the structure materials and released heat of the neutron flux, the winding pack experiences irregular temperature fluctuations due to plasma disruptions. Such disruptions effect the operation of the toroidal field coil by various effects:

- Eddy currents of approximately 13 MJ are induced in the casing of the winding pack, which leads to a temperature rise of 13 K at the inboard leg.
- The current in the TF coil can increase to values of up to 100 kA.
- Coupling losses of about 0.8 MJ are expected.

In order to withstand such enormous heat input, a minimum current sharing temperature of 5.7 K is required for the CICC at an operation current of 68 kA and a background field of 11.8 T [BM08]. First CICC measurements have shown, that this value is exceeded by minimum 0.4 K [BBC+08]. The ratio between operating and critical current at 11.8 T and 5.0 K is estimated with

$$\frac{I_{\mathrm{op}}}{I_{\mathrm{c}}} = 0.78,$$

which corresponds to a critical current of 87.2 kA. Measured n-values of the Cable in Conduit Conductor are typically around 10, depending on the twist pitch and the experienced strand damage during the in-field measurement [BBC+08, BBD11]. An increased n-value is only achieved with longer twist pitches and smaller helium voids[BBC+08].

3.3 DEMO TF coil design parameters

A final design of the DEMO reactor, including the toroidal field coil parameters, does not yet exist. Instead, the reactor design is actually developed in an iterative process, which combines physical and engineering parameters, estimated, proposed and checked for feasibility by various participating research institutes. Core of this iterative process is the system code PROCESS [KK16] run at the Culham Centre for Fusion Energy, Oxfordshire, which combines the defined physical and engineering boundary conditions to calculate the optimum parameters of the European DEMOnstration Power Plant. In order to proceed with the development of a conductor for a DEMO toroidal field coil, it was necessary to stick to one PROCESS data output. For following work, the PROCESS output of the July 2012 [Kem12, MBKF15] build was taken as basis.

The DEMO magnet coil architecture consists of 16 circular arranged toroidal field coils with a total, stored energy of 144 GJ. The entire current per coil integrated of the number of windings is 19.06 MA per coil, which induces the required magnetic field of 6.85 T along the central line of the plasma torus. For the same reasons as already mentioned for the ITER coil design, the DEMO TF coil has a D-shaped form. The aspect ratio for the plasma is with 4.0 extraordinarily high compared to former fusion experiments. The estimated values for minor and major plasma radius are $a = 2.25\,\mathrm{m}$ and $R_0 = 9.0\,\mathrm{m}$. As the PROCESS output data suggested toroidal field coil dimensions without a continuous coil function and a bent inboard leg, it needed to be adjusted to a design with a straight inboard leg and a continuous coil function [GBB$^+$14, Gad16]. The resulting coil design is visualized in figure 3.12. Compared to ITER it is almost three meters higher and about one meter wider to provide space for the 1.8 times larger plasma volume. The main difference to the ITER TF coil is the axial symmetry to the horizontal line. This symmetry will most likely be broken in future development iterations, as the divertor demands design adjustments. All 16 inboard legs share a trapezoidal cross-section area and together form the centre column of the reactor chamber.

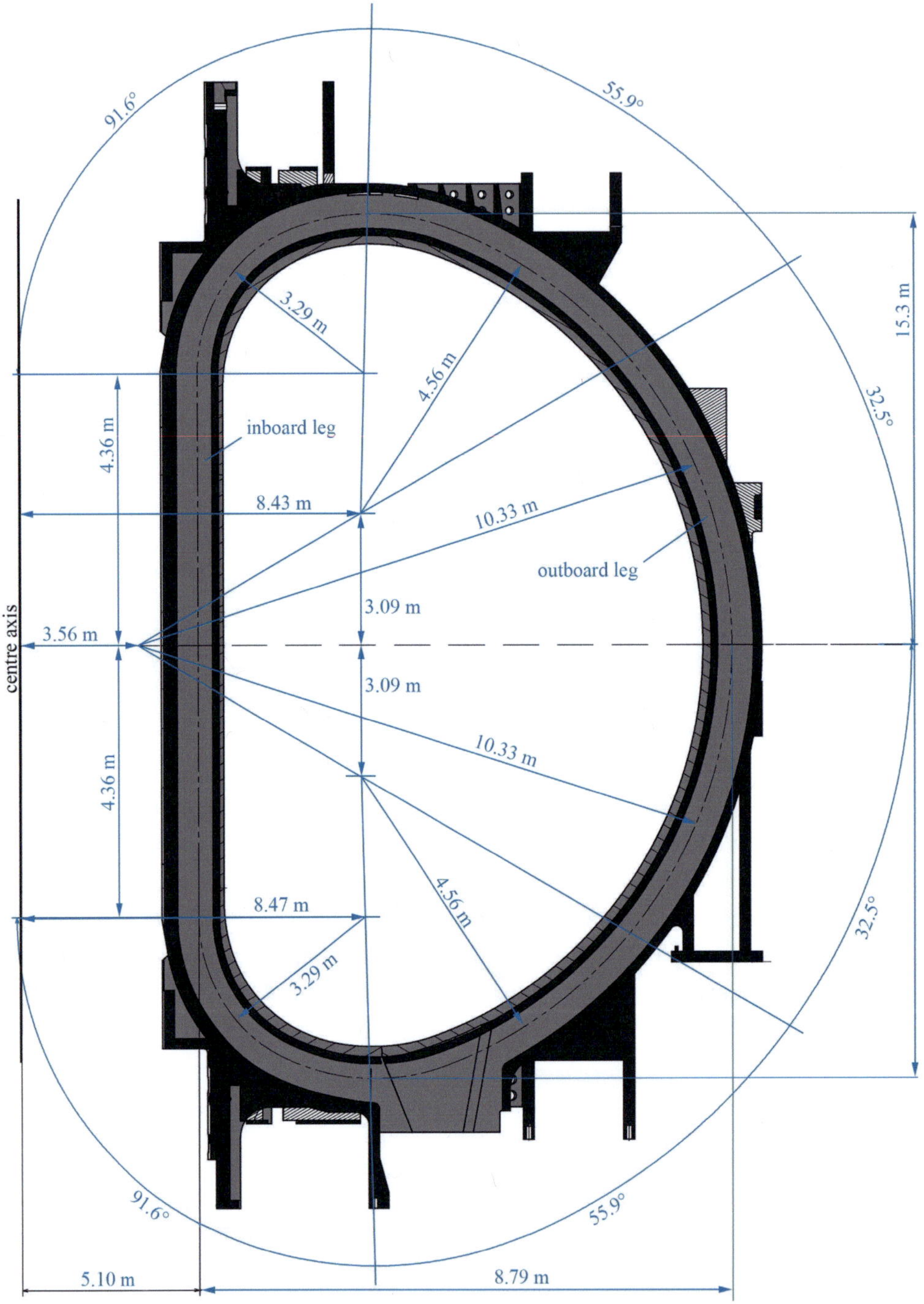

Figure 3.12: Dimensions of the DEMO toroidal field coil according to the adjusted DEMO July 2012 design [GBB+14, Gad16]. Although the dimension are correctly denominated, the drawing is not true to scale.

3.3.1 Winding pack

Actual calculations suggest a winding pack area of 1.10 m^2 [HGF+16, Gad16] with 384 coil windings. With an entire current per coil of 19.06 MA, this corresponds to a current of approximately 50 kA per conductor winding. Calculations from [Gad16, HGF+16] suggest 18 pancake windings. The four outermost pancakes consist of 19 windings and the 14 inner pancakes of 22 windings each. The pancakes are separated by a 3 mm thick pancake and a 10 mm thick ground insulation. A cross-section view of the winding pack is shown in figure 3.13.

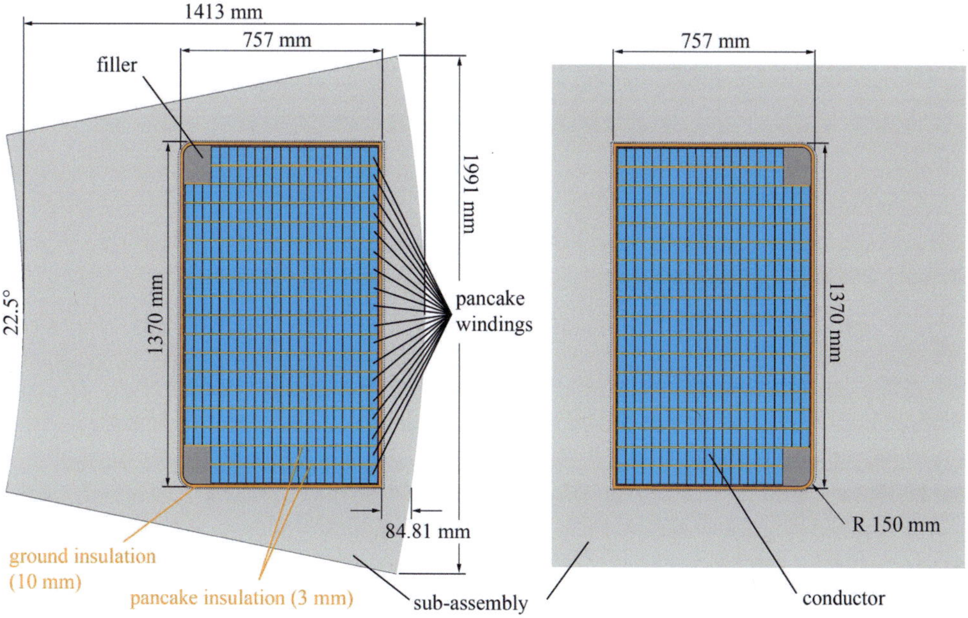

Figure 3.13: DEMO winding pack for inboard (left) and outboard leg (right).

The exact winding pack design for each of the 384 windings is given in figure 3.14. These dimension are taken as basis for calculations in chapter 7 and, depending on the cable design, it is recommended to increase the conduit thickness to 6.4 mm.

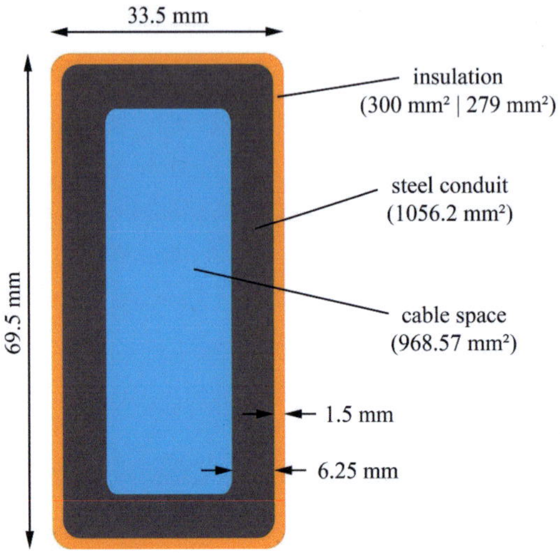

Figure 3.14: Dimensions of the DEMO cable conduits.

3.3.2 Longitudinal strain

During coil operation, induced Lorentz forces try to pull the conductors apart. This effect is called hoop stress. In case of large scale magnet coils, such as the ITER and DEMO TF coils, hoop stress mostly acts in longitudinal direction as tensile strain. For ITER the stress is estimated with 0.1% [ITE09b]. As there is no PROCESS data output on expected strain, yet, the ITER value of 0.1% is taken as reference for later strain measurements.

3.3.3 Summary

The boundary conditions for the conductor design of DEMO are the PROCESS output and theoretical requirements from discharge and Mises stress calculations were executed by [Gad16]. Table 3.3 summarizes all coil and winding pack properties and boundary conditions for ITER and for a future DEMO conductor design.

Table 3.3: Summary of coil, conductor and winding pack properties and requirements of the ITER and European DEMO toroidal field coils.

TF coil parameters	ITER	DEMO[1]
number of TF coils	18	16
total current in one TF coil (MA)	9.1	19.06
stored energy in one TF coil (GJ)	2.27	9.09
peak toroidal field at plasma (T)	5.3	6.82
peak magnetic field at TF conductor (T)	10.8	13.27
total available winding pack area (m^2)	0.49	1.10
number of conductor windings	134	384
helium inlet temperature (K)	4.5	4.5
conductor cross-section area	**ITER (CICC)**	**DEMO[1] conductor**
Nb_3Sn (mm^2)	475	-
*RE*BCO (mm^2)	-	1.82
copper stabilisation (mm^2)	524	554.9
silver (HTS) (mm^2)	-	5.48
helium void (mm^2)	322	272.2
stainless steel (jacket + HTS substrate) (mm^2)	262	1147.25
stainless steel (radial plates + jacket + casing) (mm^2)	4194	-
insulation (mm^2)	> 276	300
conductor performance	**ITER (CICC)**	**DEMO[1] conductor**
operating current I_{op} (kA)	68 at 10.8 T, 5.0 K	50 at 13.27 T, 5.0 K
safety factor (I_{op}/I_c)	0.78	0.7
critical current I_c (kA)	87.2 at 10.8 T, 5.0 K	71.4 at 13.27 T, 5.0 K
current sharing temperature T_c (K)	5.7 at 10.8 T, 68 kA	16.4 at 13.27 T, 50 kA
n-value	~ 10	to be evaluated

[1] DEMO July 2012 design

4 Development of soldered joints and contacts

As already discussed in chapter 3.1.4, 2G high temperature superconductors are manufactured in a layered structure. With the state-of-the-art manufacturing methods, the longest single piece length achieved to date is several hundred meters [Sup15][Ame15]. A toroidal field coil for DEMO with a single pancake or layer winding requires a cable length of about 700 m. For double pancake winding it is even 1.5 km. Therefore it will be necessary to adapt the length of the manufactured superconductors to the desired cable length. In this chapter different solutions for cable joints are analysed with focus on single tape solder joints. Aspects which affect the solder joint quality are discussed. Experiments on the reproducibility of solder joints are evaluated and afterwards transferred to full-size DEMO cable contacts.

4.1 Requirements

The following discussion on joint design focuses on inner joints, which are explicitly required to extend the cable length. Although this excludes joints for pancake and layer winding it will be denominated as *cable joints* in the following discussion.

Inner cable joints are feasible on three levels: either they connect complete cable sections (butt joint), smaller single strands (splicing) or simply single tapes. In consequence, the joint design depends on the desired level of connection. Each of these three connection types is visualized in figure 4.1.

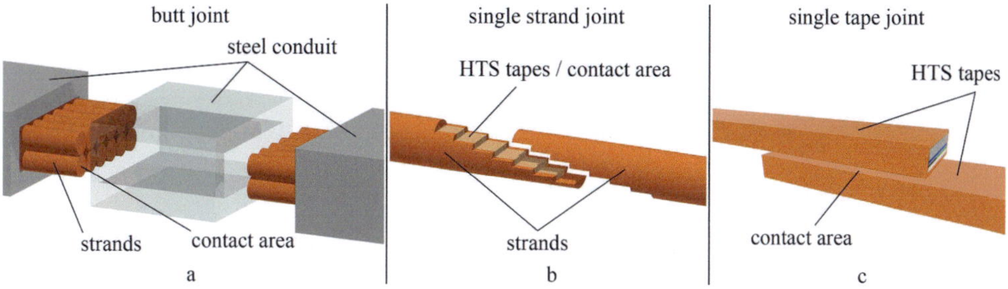

Figure 4.1: Schematic of a cable butt joint, a single strand joint and a single tape joint.

Butt joints (figure 4.1 a) are not applicable for superconducting DEMO cables. The cross section-area of a butt-joint is limited by the cable dimensions already discussed in chapter 3.3.

This includes a cable width of 69.5 mm and a cable height of 33.5 mm which corresponds to a cross-section area of 2328 mm^2. Since the cable cross-section is completely filled with superconductor strands, helium void, the stainless steel jacket and the insulation, there is no space left to guide the current through a copper bypass. Without such bypass it is only possible to directly connect single strands with each other.

This leads to single strand joints, where single strands are spliced and connected individually. Each single strand joint can be described by a step function of n steps, where n corresponds to the number of HTS tapes per strand (figure 4.1 b). Naturally, the minimum length of each step needs to match the conductor transition length, to keep the ohmic heating to a minimum level. Although such a joint technique is quite feasible, it is not the optimum solution for a multi strand cable. It would be necessary either to manufacture the strand with a precise displacement for each tape to realise the step at the strand ends, or to mill the steps afterwards, taking the risk of tape damage. A strand joint is an accumulation of single tape joints which dissipates heat and favours the spontaneous loss of superconductivity, referred to as quench.

The optimum connection is a joint already performed on single tape level to adapt the single tape length to the desired cable length (figure 4.1 c). This allows to manufacture a full-length cable strand without further connections on strand or even cable level. Single tape joints are less complex than single strand joints and therefore more feasible. Furthermore, an accumulation of localized hot spots can be avoided: by shifting the single tapes against each other in longitudinal direction during the manufacture process, the single tape joints, and therefore the dissipated heat, can be distributed over the entire strand length. The ratio of the local heat capacity, provided by the joint surrounding material, to the number of joints increases. In case of a quench, which likely occurs at hot spots, this is the favourable situation. In consequence, single tape joints are the method of choice to reach the full-size DEMO cable length. Hence, the following work concentrates exclusively on single tape joints, their feasibility and their transferability to cable terminations.

4.2 Joint definitions

Due to the layered structure of the *RE*BCO tapes, three kinds of joints can be distinguished: the face-to-face, the back-to-back and the face-to-back lap joint. The first two can be seen in figure 4.2.

The optimum configuration for a single tape joint is the face-to-face lap joint with both *RE*BCO layers facing each other. Compared to a back-to-back or face-to-back lap joint, where the current has to pass the high ohmic stainless steel substrate layer, a face-to-face lap joint has a much lower ohmic resistance. In the following chapters only face-to-face lap joints will be

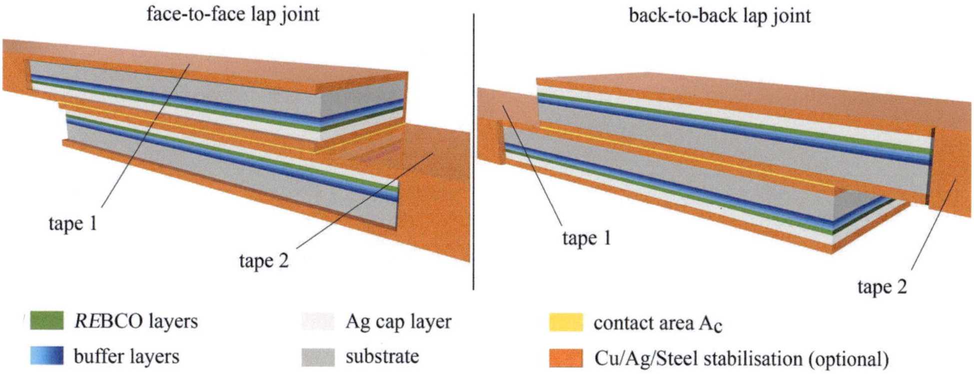

Figure 4.2: Cutaway drawing of a face-to-face and a back-to-back lap joint made of two HTS tapes.

discussed. Back-to-back and face-to-back lap joints must be avoided.

Furthermore, the investigation focuses on soft soldering, which accounts for soldering temperatures below 450 °C [FS08]. Mechanical contacting methods or diffusion annealing were not considered in the evaluation for two reasons: while mechanical contacts come with high ohmic resistances, diffusion annealing is executed at around 200 K below the solidus temperature of the participating constituents. Considering a copper stabilisation, temperatures of about 900 °C would be necessary. At such temperatures oxygen leaves the superconducting Cu-O layer in the *RE*BCO lattice which destroys the superconductor.

Official recommendations by industrial manufacturers are temperature limits of 200 °C for Zr-doped $REBa_2Cu_3O_x$– [Sup14] and 175 °C for Dy-doped $Y(Dy_{0.5})Ba_2Cu_3O_{7\text{-}\delta}$ superconductors [RVZ+04][AMS].

The following work on soft solder joints consists of statistical analysis of preceding publications, evaluations and verifications of the most promising solutions to satisfy industrial feasibility. The criteria taken as basis are

- low ohmic resistance,
- reproducibility,
- no tape degradation,
- and homogeneous current input for all tapes.

The huge range of publications on HTS soft solder joints is summarised in the data base annex **??**. To be able to compare and evaluate the contact resistances, definitions are mandatory. Therefore the intrinsic resistance R_{in} and the contact resistance R_c [VSW02] are defined. The intrinsic resistance R_{in} denotes the resistances of each material layer between the *RE*BCO layers. In the optimum case of a face-to-face lap joint, this is mostly the silver and copper stabilisation. The contact resistance R_{c} represents the microscopic contact between the crystal lattices of two contacting materials. Both resistances are visualized in figure 4.3.

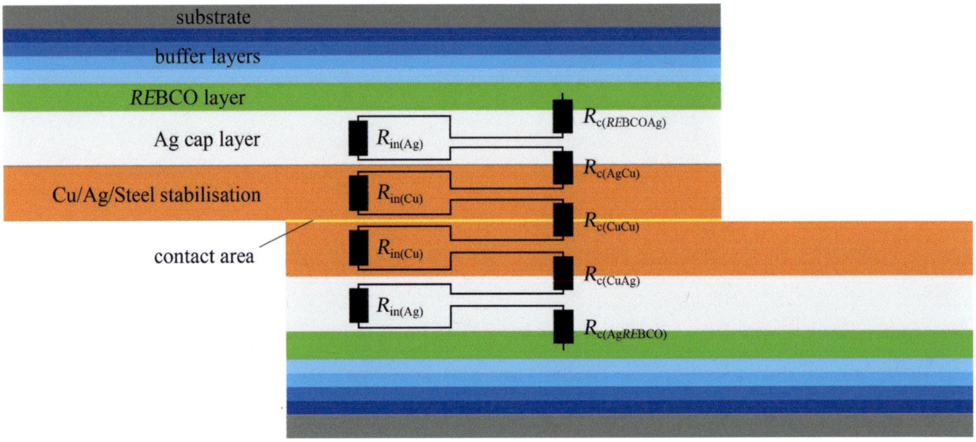

Figure 4.3: Intrinsic joint resistances R_{in} and contact resistances R_{c} for an HTS face-to-face lap joint.

Since technical superconductors are self-contained systems, we define the joint resistance R_{j} for a face-to-face lap joint by

$$R_{\text{j}} = \sum R_{\text{in}} + \sum R_{\text{c}}.$$

Naturally it is only meaningful to compare the joint resistance R_{j} under consideration of the contact area A_{c}. Therefore we define the specific joint resistance

$$r_{\text{js}} = R_{\text{j}} \cdot A_{\text{c}}.$$

These definitions will be used to compare and evaluate results from the data base annex **??**.

4.3 Impact of joint specific properties

4.3.1 Soft solder

The soft solder type is an influencing factor on the specific joint resistance. To get an overview of achieved soft solder lap joints within the last years a literature research was foregoing. All results are summarized in figure 4.4, sorted by soft solder.

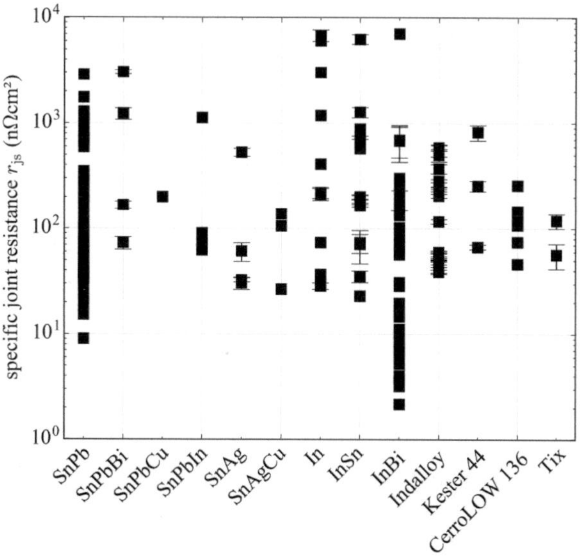

Figure 4.4: Specific joint resistances r_{js} of all collected publications on soft solder contacts between 2G high temperature superconductors, sorted by solder. Error bars indicate the standard uncertainty u. The source data of the plot can be found in table ?? in the annex A.1.

Depending on the different preference of the investigated soft solders, the sample size varies accordingly. Regarding the sample size, the most common solders are SnPb, InSn and InBi. The minimal specific joint resistance r_j was achieved with InBi with less than 10 nΩcm². Indium, InBi and InSn are the soft solders with the largest spread of almost three orders of magnitude between the maximum and the minimum measured specific joint resistances.

To investigate the impact of the intrinsic resistivity of each soft solder type on the overall specific joint resistance, it is essential to compare their physical properties. Table 4.1 summarizes relevant physical properties of the most common soft solders.

With 15 $\mu\Omega$cm InBi is the soft solder with the highest intrinsic resistivity at 77 K. That corresponds to an intrinsic resistance of 1.5 nΩcm^2 for a solder layer thickness t_s of 1 μm. For the other soft solders the intrinsic resistance at the same solder layer thickness is even less. The intrinsic resistance R_i of the soft solders can directly influence the joint resistance R_j. Therefore, it is recommended to keep the layer thickness t_s to a minimum level to reduce any possible intrinsic contribution. A plausible way to do so, is to apply pressure during the soldering process, which will be discussed in section 4.5.1.

The benefit of using soft solders with a high intrinsic resistance such as InSn or InBi derives from their low liquidus temperature. As already mentioned in section 4.1 the maximal recommended soldering temperature is 200 °C and 175 °C, before oxygen starts to diffuse from

Table 4.1: Physical properties of common soft solders. The intrinsic resistance was calculated for a soft solder layer thickness of 1 μm.

	composition (m%)	T_l (°C)	ρ(77 K) ($\mu\Omega$cm)	r_{in}(77 K) (nΩcm^2)	data from
InBi	In66Bi33	72	15	1.5	[SDO10, BBW14]
In	In100	157	1.6	0.16	[ZDHG11, Ind08]
InSn	In52Sn48	118	9	0.9	[ZDHG11, BBW14, Ind08]
SnPb	Sn60Pb40	191	2.5	0.25	[TAB+12, BBW14, Ind08]
SnPb	Sn63Pb37	183	2.5	0.25	[LHS+11, BBW14, Ind08]
SnAg	Sn96.5Ag3.5	211	2.5	0.25	[DZG+10, BBW14, Ind08]

the *RE*BCO layer. With liquidus temperatures of 118°C and 72 °C, both, InSn and InBi allow for long and extensive soldering periods, without taking the risk of damaging the superconductor.

4.3.2 Flux

Flux helps to improve the contact resistance R_c during the soldering process. Its beneficial properties derive from its composition which includes activators, vehicles and solvents.

- Activators support the soldering process by dissolving oxidised metals and favouring the alloying process by increasing the iron concentration.
- Vehicles remove the reaction residues of the soldering process and protect the hot metal surface against further oxidation.
- Solvents are not necessarily used, but allow for a better deposition of the chemical compounds.

Working with LTS and BSCCO superconductors for a very long time, the experience at our institute has shown, that best results could be achieved using Spirflux330® or colophony dissolved in isopropyl alcohol. The physical properties are listed in table 4.2.

In contrast to LTS and BSCCO superconductors, the superconducting layer in 2G HTS superconductors is, depending on the cable design or the solder joint, in direct contact to the flux. Therefore, Spirflux330® and colophony solved in isopropyl alcohol were investigated for degradation effects on HTS wire manufactured by SuperPower Inc. Twenty samples were prepared in total. Each sample had an original tape width of 12 mm, which was reduced when the edges were removed, to provide for direct contact between the *RE*BCO layer and the flux. To calculate the relative degradation, the critical current I_c and the n-value n were measured,

Table 4.2: Parameters of Spirflux330® and colophony.

	Spirflux330®	colophony dissolved in isopropyl alcohol
flux type	organic	organic
ph	2	7, neutral
composition	90% distilled water < 10% glutamic acid salt < 5% urea and thickening agent < 0.5% dye, reddish	66.7% isopropyl alcohol 23.3% resin 6.7% turpentine 3.3% water

before each sample was put into a flux bath for a different period of time. Afterwards, the samples were cleaned with purified water to remove any residues and were measured again. The sample size for each measurement is two. Figure 4.5 shows the time-dependant influence of the fluxes on the *RE*BCO superconductor.

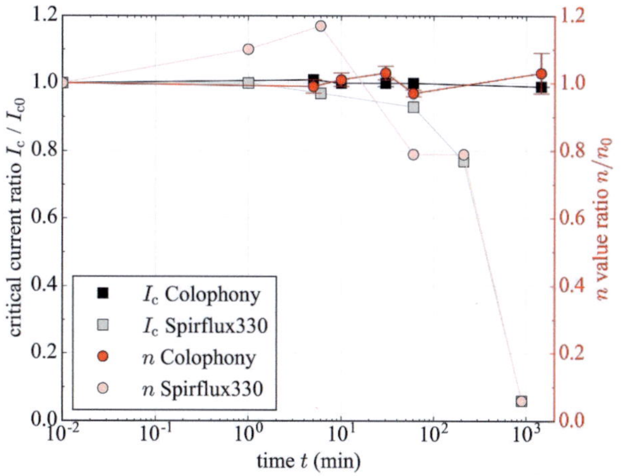

Figure 4.5: Critical current and n-value dependency of *RE*BCO on colophony and Spirflux330®. Both reference values I_{c0} and n_0 were measured for each tape before the flux treatment. Error bars indicate the standard uncertainty u for a sample size of 2. The solid lines are not physical, but serve as visual guide.

Spirflux330® impairs the *RE*BCO layer. After few minutes, the superconductor starts degrading irreversibly. Leaving the superconductor for 15 h inside the flux results in a complete degradation. A superconducting transition could not be measured anymore. With a ph of 2, Spirflux330® is acid and attacks the *RE*BCO lattice and the metallic stabilisation. Since it is not possible to ensure that no chemical residues, that might creep into the *RE*BCO layer, are left, Spirflux330® is not even qualified for short soldering periods of 2G HTS. The way more promising flux is colophony dissolved in isopropyl. Even after 24 h, no impairment of the current transport capability could be measured. This does not only guarantee for successful short-term soldering, but also for thoroughly, time-consuming soldering on complex HTS cable structures.

4.3.3 Contact area

The most direct contribution to the joint resistance R_j is the area of contact A_c. With increase of the area of contact, the absolute joint resistance drops. Such behaviour derives from the relation

$$R = \frac{\rho}{A_c} \cdot l,$$

where ρ is the electrical resistivity. The joint resistance drops in the expected $\frac{1}{A_c}$ proportionality as can be seen in figure 4.6.

For the specific joint resistance r_{js}, which is mathematically not dependant on the contact area A_c, a constant level is expected. However, when looking closely at figure 4.6, a downtrend of the specific joint resistance is visible for Indalloy, InBi and InSn. This decrease is less a physical effect than a systematic error. The tape edges of coated conductors are not sharp, but rounded, diminishing the quality of the total contact area. Therefore solder joints with a high ratio of tape edges compared to the contact area are of lower quality. With increase of the contact area A_c, the ratio of tape edges to contact area decreases in favour of the joint quality.

Fluctuations in the specific joint resistance as seen for SnPb lie within typical standard uncertainties for soft solder joints and therefore are a matter of systematic and statistical error.

Another aspect in regard to the specific joint resistance is the transfer length λ, which defines the length, the current needs to fully couple into the *RE*BCO layer. For 6 mm wide superconductor tapes, the length was measured $\lambda \ll 1$ cm [PBL06, BAC⁺15]. In case of the evaluated data, the contact length exceeds the coupling length by far. Therefore, the contribution of the coupling can be neglected.

4.4 Impact of conductor specifications

4.4.1 Production batch

The fact that the tape quality respectively the critical current I_c fluctuates over the produced tape length is well known. Therefore it seems likely, that the tape quality might have an effect on the contact resistance R_c between the intrinsic layers and therefore on the specific joint resistance r_j of a soft solder joint.

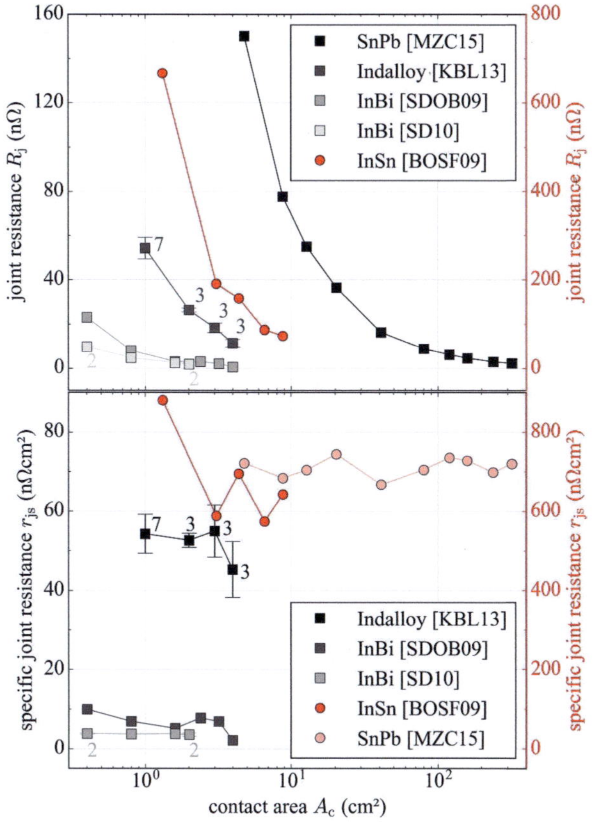

Figure 4.6: Joint resistance R_j and specific joint resistance r_{js} over the contact area A_c per meter conductor length. While the joint resistance decreases with a $\frac{1}{A_c}$ proportionality, the specific joint resistance remains constant. The sample size is given by numbers. Solid lines are not physical, but indicate the physical trend. Error bars indicate the standard uncertainty u and numbers the sample size.

During the manufacturing process, the superconductor of several hundred meters length is cut into shorter sections and wound on spools. Each spool is denominated with a spool ID and the suffix *MX*, where *X* is a natural number between 1 and 3, marking the production line. Furthermore, 4 mm wide single tapes are cut from raw tapes with a width of 12 mm [Sup12]. Each of the three resulting single tapes is denominated after its slit position. The denominations are *front-slit (FS), middle-slit (MS)* and *back-slit (BS).*

The influence of the production batch on solder joints was investigated by Kim, Y. et al.,for SCS4050-AP wires from SuperPower Inc. Both, pre-tinned and non-pre-tinned samples were taken into account. As can be seen in figure 4.7, the specific joint resistance r_j is highly depending on the spool ID.

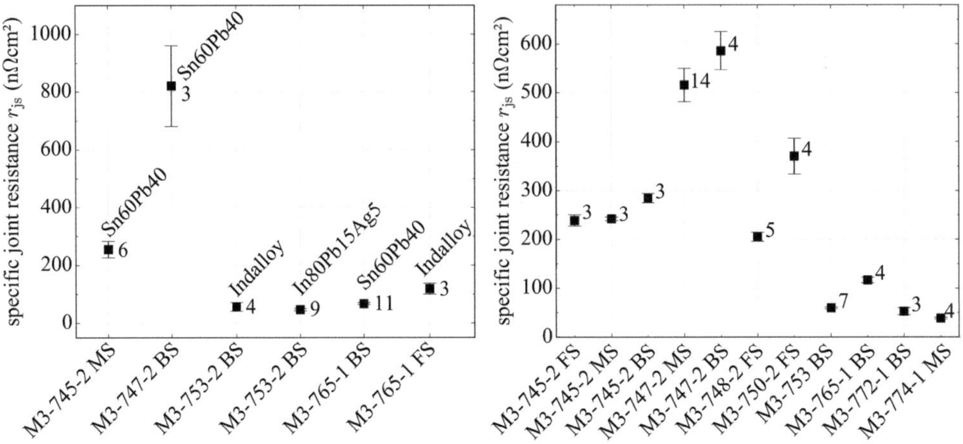

Figure 4.7: The production batch is highly influencing the specific joint resistance r_j for both, pre-tinned and non-pre-tinned solder joints. Although the same soldering procedure is performed for all measurements, the results deviate for different spool_IDs. Error bars indicate the standard uncertainty u and numbers the sample size. Data from [KBL$^+$13].

Different soft solders were used for pre-tinned and Indium was used for non-pre-tinned samples. Solder joints made with superconductors of the same spool ID such as M3-745-2, M3747-2 and M3-753-2 share similar specific joint resistance values for different slits (*FS*, *MS*, *BS*). Even the soft solder type has less impact on the specific joint resistance r_{js} than the spool ID, as can be seen by comparing M3-753-2 with M3-747-2.

No correlation between the critical current quality of the tape and the solder joint quality was found. Therefore, it is not possible to derive the joint quality from the current carrying capability of the superconductor.

Following two explanations are assumed:

- the intrinsic resistance R_{in} between the superconducting *RE*BCO layer and the silver and copper layer is highly different for each spool ID.
- the intrinsic resistance R_{in} of each spool ID is differently susceptible to the soldering process.

Both explanations need further investigations in cooperation with the manufacturer and will not be discussed at this point.

Concluding, it can be said, that for the investigated samples, the specific joint resistance r_j scales with the spool ID. Therefore, the manufacturing process has a high impact on the resulting contact resistance. To reduce such impact, the contact area A_c should be as large as possible. After all, it is not the specific joint resistance r_{js}, but the absolute joint resistance R_j of each single tape, which guarantees for a homogeneous current distribution within the cable. The discussion on the homogeneous current input will be continued in chapter 6.1.1.

4.4.2 Electrical stabilisation

As described in subsection 3.1.4, most HTS tapes are electroplated with a metal stabilisation made of Cu, brass or stainless steel (SS). Brass and stainless steel are less common and highly dependant on the application. Nonetheless, the impact of any of these metal stabilisations on soft solder joints is important.

Regarding the joint resistance R_j, both, the intrinsic resistance R_{in} and the contact resistance R_c have to be considered. The intrinsic resistance is directly dependant on the given electrical resistivity ρ which highly defers for Cu, brass and SS. Its contribution can be easily calculated and removed from the measured joint resistance. The result can be seen in figure 4.8. Round markers represent the specific joint resistance r_j including the intrinsic resistance R_{in}, while square markers represent the specific joint resistances after subtracting the intrinsic contribution. Therefore, the specific joint resistance mainly derives from the contact resistance R_c between the metal stabilisation and the silver layer respectively the metal stabilisation and the solder layer.

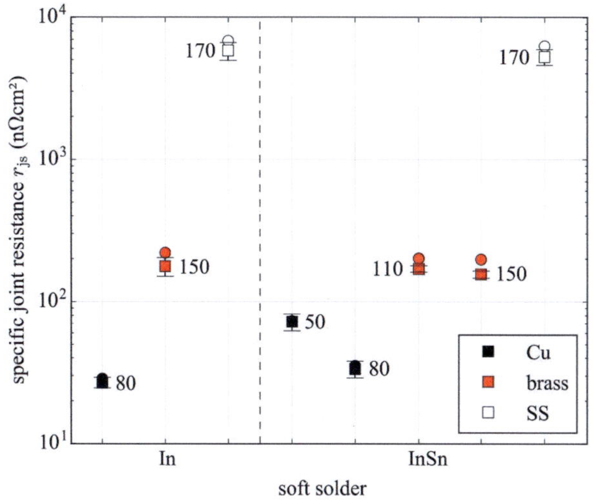

Figure 4.8: Specific joint resistance r_{js} for different electroplated metal stabilisations. Circles give the value before subtracting ohmic contributions due to the intrinsic resistance R_{in} of the stabilisation material. Squares give the corrected data including error bars which indicate the standard uncertainty u. Numbers give the stabilizer thickness in μm. The sample size is 6 for each data point. Data from [ZDHG11, DZG$^+$10].

Comparing the joint specific resistance for Cu, brass and stainless steel stabilisations, copper is the one which results in the lowest resistance, independent of the soft solder type. Brass follows the lead with an order of magnitude higher specific joint resistance. Stainless steel concludes with a specific joint resistance of almost 10 $\mu\Omega$. Therefore a Cu-stabilisation is a mandatory requirement to achieve solder joints with low specific joint resistances.

Although such statement favours the usage of Cu-stabilized HTS tapes in full-size cables for a toroidal field magnet coil, it is not the only decisive argument. Copper is the HTS tape stabilisation of choice for three reasons:

- it favours solder joints with low specific joint resistances,
- it has a low ohmic electrical resistivity, which is necessary to transport current during a superconductor quench,
- and it has a high thermal conductivity, which is prerequisite for a good heat exchange between the *RE*BCO layer and the cooling medium.

4.5 Mechanical properties

4.5.1 Transversal pressure

As already mentioned in subsection 4.3.1, transversally applied pressure during the soldering process is a possibility to reduce the thickness of the solder layer [SDOB09, SDO10, CF12]. The influence of applied pressure during the soldering process is illustrated in figure 4.9.

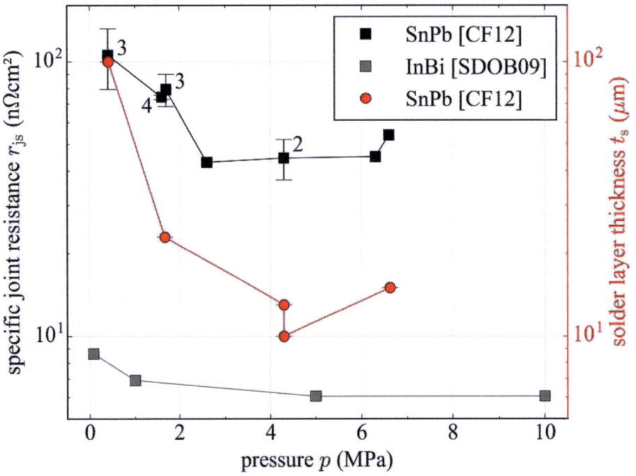

Figure 4.9: Specific joint resistance r_{js} and solder layer thickness t_s, depending on the applied pressure during soldering. The solder layer thickness was not measured for all tapes, but for random samples. Error bars indicate the standard uncertainty u and numbers the sample size. Data from [CF12, SDOB09].

If the applied pressure is increased, the resulting specific joint resistance r_{js} drops. For solder joints made with SnPb it is possible to reduce the specific joint resistance value by almost

50% for pressures above 4 MPa. Solder joints made with InBi could be optimized by 30% for pressures above 5 MPa. The decrease of the specific joint resistance correlates with the solder layer thickness. High transversal pressure applied during the soldering process reduces the solder layer thickness t_s significantly. In case of the SnPb solder joint, a pressure of more than 4 MPa leads to a reduction of the solder layer thickness by more than 80%. This implicates a reduction of the intrinsic resistance of the solder layer of same magnitude. For the SnPb measurement, the reduction of the intrinsic resistance at 4 MPa accounts for $21.25\,n\Omega cm^2$. As this value does not cover the 50% reduction, the contact resistance R_c seems to improve as well.

Concluding, the recommendation is to apply transversal pressure of more than 5 MPa for face-to-face lap- or bridge joints.

4.5.2 Longitudinal strain

Circular conductors inside a magnet experience various types of stress. This includes coil winding stress, thermal stress and hoop stress due to the Lorentz forces driving the conductors apart in radial direction (3.2.3). While thermal stress and the coil winding stress can be regulated during manufacturing and by the cool down rate of the toroidal field coil, hoop stress is unavoidable during operation. Hoop stress acts circumferentially on the superconductor cable and therefore, due to the large coil radius, in longitudinal direction on the superconductor tapes and the soft solder joints.

Investigations on longitudinal strain of soft solder joints have shown, that the limiting factor of a soldered lap-joint under longitudinal strain is not the joint, but the strain experienced by the HTS tapes right before and after the joint [DZG+10, SDO10, IIA+14, LZL+14]. The resistance of the lap-joint remains constant during applied strain, although the critical current drops as soon as the HTS tapes come to their mechanical limit. Therefore, by mechanically stabilising the HTS tapes against hoop stress in toroidal field coils, which will be further discussed in chapter 7, lap-joints are already covered and do not require a special treatment.

4.6 Nuclear activation

In a fusion reactor where electric energy is generated from kinetic energy of emitted, free neutrons, nuclear activation is unavoidable. Nuclear activation is not only limited to the blankets, which are in direct contact with the neutron flux, but also concerns the superconducting magnets. Fusion power station is intended to be fully recyclable after 100 years. This includes a decrease of the on-contact dose rate D/t to a level of 0.01 Sv/h. Dose rates for constituent materials in *RE*BCO after being activated for five years with the first wall dose (FWD) of the ITER reactor were already discussed in [Bar13]. With the use of soft solder joints, the first wall dose also activates the constituent elements of the solder alloys. Except for In, the decrease of the corresponding on-contact dose rates can be seen in figure 4.10.

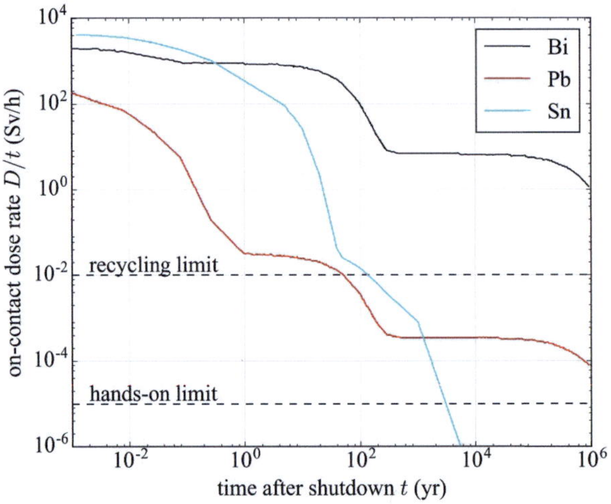

Figure 4.10: Time dependant decrease of the on-contact dose rates of constituent elements in soft solder joints after a nuclear activation of five years by ITER first wall dose (FWD). Data from [Sim16].

Radioactivity measurements in a D-T fast neutron field have shown that the expected half-life is 54.2 min for ^{116m}In and 4.49 h for ^{115m}In [KAB+95]. Therefore, the hands on limit is reached within a short time. Several activated Pb and Sn isotopes have longer half-lives, but the recycling limit is still reached after 100 years [Sim16]. For Bi the recycling limit is not even reached after more than 10^6 years [Sim16]. In order to stick to the recycling limit of 100 years, it is necessary to avoid Bi in soft solder joints.

4.7 Reproducibility

Since all hitherto discussed factors influence the joint resistance of 2G HTS joints, reproducibility becomes very important. In order to get an idea of the feasibility, solder joints for three soft solders were manufactured, with a sample size of ten per solder. Table 4.3 gives an overview of the considered soft solders. Both, InBi and InSn were selected for their low soldering temperatures and SnPb was selected for its wide acceptance within the HTS community.

Table 4.3: Soft solders tested for reproducibility with the maximum temperature reached during the soldering process.

solder	composition (m%)	T_l (°C)	T_{max} (°C)	manufacturer
InBi	In66.3Bi33.7	72	95	Chemet
InSn	In50Sn50	125	150	Chemet
SnPb	Sn60Pb40	191	210	Stannol

Before creating the lap-joints, each tape was preprocessed. This included

- removing the oxygen layer with a scouring pad (up to 20 iterations),
- cleaning the surface from dust and grease with isopropyl alcohol,
- and pre-tin each tape on a hot plate.

After the pre-tinning, the flux on colophony basis was washed off with isopropyl alcohol. The subsequent soldering was executed under a constant pressure of 6 MPa with fully monitored heating and cooling sequences.

Figure 4.11 shows the results of the manufactured solder joints compared to the joints from data base **??**. Each black marker represents the mean value of a single study with the corresponding standard uncertainty u. Red markers represent own measurements of produced joints with the corresponding standard uncertainty.

For SnPb the mean reproduced specific joint resistance is $47.73 \pm 5.06\,\mathrm{n\Omega cm^2}$ with a sample size of 8, for InSn at $102.68 \pm 5.25\,\mathrm{n\Omega cm^2}$ with a sample size of 10 and for InBi at $160.68 \pm 17.77\,\mathrm{n\Omega cm^2}$ with a sample size of 9. These results lie below the average value of the compared publications, summarized in table **??**.

4.8 Conclusion

Single tape joints are the recommended solution to extend the superconductor cable length for three reasons. The extension on tape level allows

- to extend the the tapes without splicing,
- to manufacture a single cable of the desired cable length
- and to distribute the resistive hot spots in longitudinal direction over the entire cable.

Data on soldering single tape joints in form of face-to-face lap joints have been summarised, analysed and evaluated in regard to the applicability in full-size DEMO TF coil fusion cables.

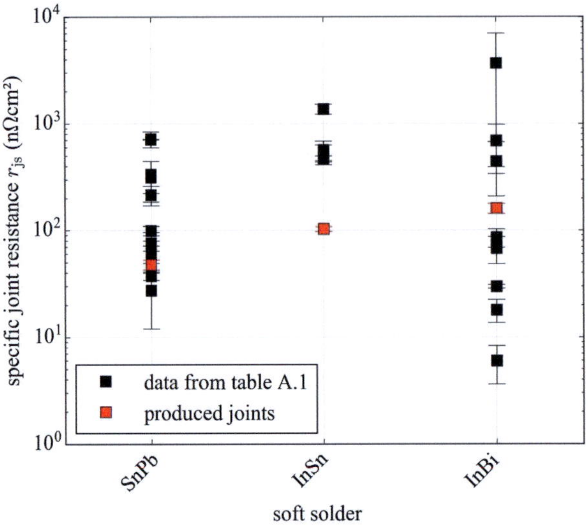

Figure 4.11: Specific joint resistances r_{js} of reproduced joints compared to data from collected publications on soft solder contacts between 2G high temperature superconductors (table **??**). Error bars represent the standard uncertainty u. The source data of the black markers can be found in in the annex A.1.

It was shown that single tape joints are highly reproducible and can reach specific joint resistances of 47.73 ± 5.06 nΩcm² for SnPb, 102.68 ± 5.25 nΩcm² for InSn and 160.68 ± 17.77 nΩcm² for InBi.

The area of contact A_c is highly influencing the absolute joint resistance R_j and should therefore be as large as possible as long as it is within the provided spatial constraints.

Differences in the joint resistance depending on the production batch can neither be neglected nor reduced. So far these random deviations are side-effects of the production process, which are further investigated by the manufacturers.

Considering the electro-mechanical stabilisation of the HTS tapes, copper is the material of choice. Its specific joint resistance r_{js} is very low compared to brass and stainless steel, it has a low ohmic electrical resistivity, which is necessary to transport current during a superconductor quench and it comes with a high thermal conductivity, which is prerequisite for a good heat exchange between the *RE*BCO layer and the cooling medium.

During the soldering process a transversal pressure of minimum 5 MPa is recommended. The applied pressure reduces the solder layer thickness t_s and therefore its intrinsic contribution to the absolute joint resistance, significantly.

Flux helps to improve the contact between the solder joint constituents. Acid flux, however, destroys the HTS tapes. Therefore, only flux based on colophony resin dissolved in isopropyl alcohol should be used.

Longitudinal strain, as experienced during hoop stress, does not impair lap joints. Instead, it is the HTS tapes which need to be protected from longitudinal strain.

The soft solder constituents Pb, Sn and In satisfy the recycling limit after nuclear activation, considering the first wall dose of the ITER experiment. For the fact, that Bi solder does not satisfy the recycling limit after nuclear activation and becomes brittle at cryogenic temperatures, ongoing work was only executed with SnPb and InSn. The voltage taps which are not in direct contact with the superconductor were soldered with miniscule amounts of SnPb. Contacts and terminations were mostly soldered with the low temperature solder InSn, which has following benefits: it is not as toxic as SnPb and allows for very long soldering periods without damaging the *RE*BCO layer.

The short summary on the three soft solders is given in table 4.4.

Table 4.4: Soft solders investigated for fusion applications.

solder	InBi	InSn	SnPb
composition (m%)	In66.3Bi33.7	In50Sn50	Sn60Pb40
T_l (°C)	72	125	191
T_{max} (°C)	95	150	210
average r_{js} (nΩcm²)	160.68 ± 17.77	102.68 ± 5.25	47.73 ± 5.06
recycling limit (yr)	> 1,000,000	~ 100	~ 100

5 Preparation of the experimental set-up for characterization of HTS cables

The experimental characterisation of HTS tapes, strands and cables is fundamental to develop a full-size cable for the fusion DEMOnstration Power Plant. Such characterisation comprises the measurement of the critical current I_c under various external boundary conditions, as they will be expected during operation of the DEMO fusion device.

This includes an expected peak magnetic background field of 13.27 T [Gad16], hoop-stress on the coil resulting in a maximum expected longitudinal strain of 0.2% [ITE06], and an operation temperature of 4.2 K with a temperature margin of 11.9 K [HGF+16]. In order to test these boundary conditions, two already existing experiments, the so called FBI (*F* tensile force, *B* magnetic field, *I* current) and the FBI-T (*T* temperature) measurement facilities, which were originally used to characterise low temperature superconductors (LTS), were upgraded and optimized for HTS measurements.

5.1 Single wire FBI measurement facility

The purpose of the single wire FBI measurement facility is the characterisation of single LTS wires and HTS tapes. The critical current I_c and n values can be measured depending on a magnetic background field B and longitudinal strain ε.

Motivation for upgrading and operating the facility was the participation in a round robin test initiated by the National Institute of Standards and Technology in 2011. The collaboration of the participating institutes was aiming for the standardisation of superconductor measurements under consideration of longitudinally applied strain on Nb_3Sn wires as they are used in the central solenoid (CS) of the ITER experiment. After commissioning the FBI measurement facility, the LTS samples were measured successfully. Today the facility is used to investigate field dependant current-strain functions of single HTS tapes.

A cutaway view of the FBI measurement facility can be seen in figure 5.1. It consists of a cryostat, a superconducting magnet, a power supply (not seen), a tensile machine (not seen) and a sample holder. Each component will be individually discussed in the following.

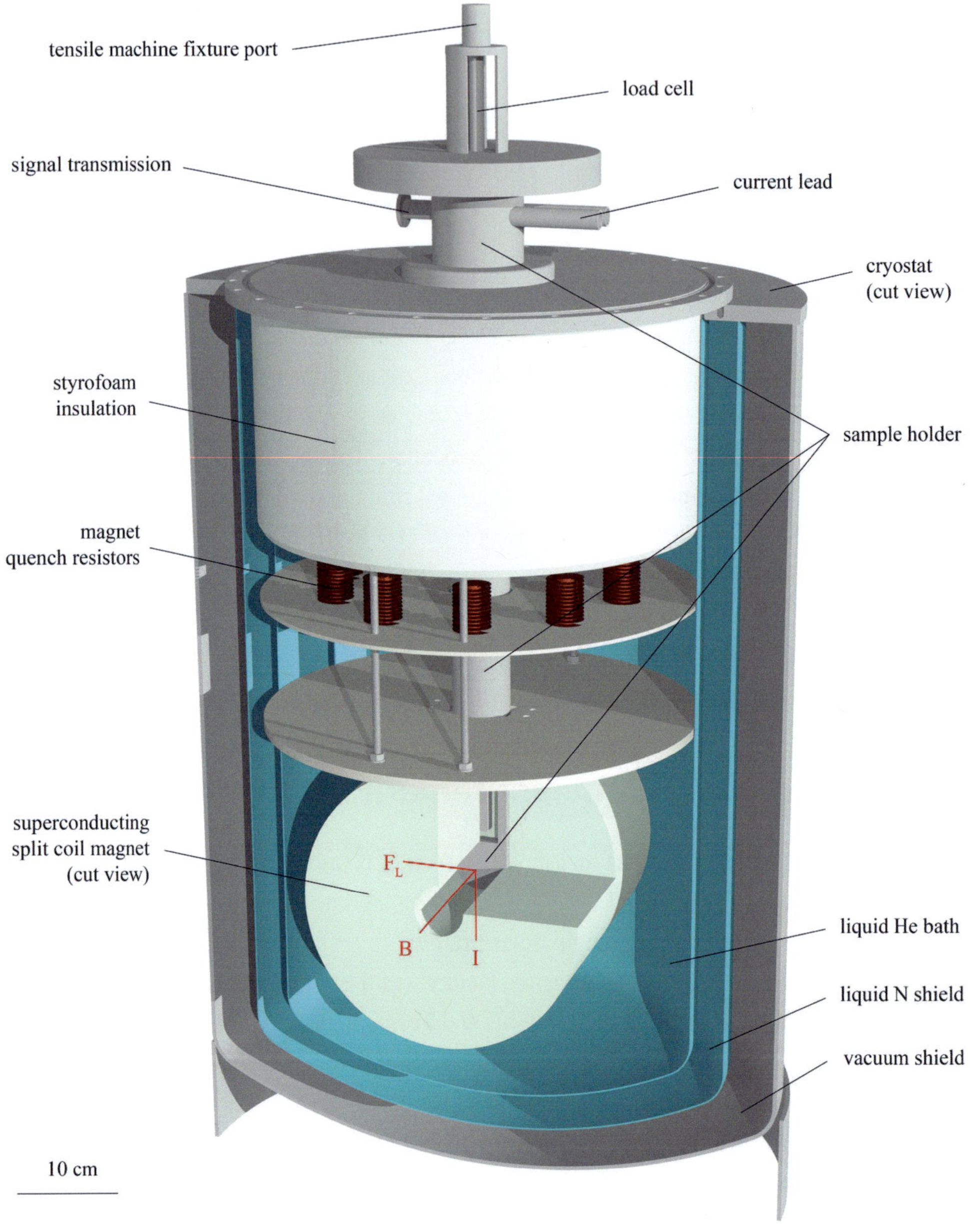

Figure 5.1: Cutaway view of the FBI measurement facility.

5.1.1 Sample holder

A sample holder is necessary to hold the superconducting sample inside the magnet centre and to stabilise it against Lorentz forces. Therefore an existing sample structure was modified and equipped with sensor cables and two CryoMak copper-beryllium extensometers to measure the current-strain functions [Nyi06].

Samples that are measured in the FBI facility are single LTS wires or single HTS tapes. Therefore it is necessary to solder the superconducting sample to the current termination and provide the necessary support against the Lorentz loads experienced during the experiments. The upper solder joint is movable in the vertical direction and directly connected to the load cell and the tensile machine fixture port on top of the sample holder (figure 5.1). The voltage taps and the extensometers are attached to the sample between the solder joints. The voltage taps are soldered with InSn soft solder and the extensometers are held in place by tension springs (figure 5.2), which allow for a fast and easy mounting.

Since the sample holder combines the sample fixture, the sample current lead and the longitudinal tensile mechanism in one single apparatus, it allows for mounting and dismounting samples while both, the cryostat and the magnet remains at cryogenic temperatures. The helium consumption and the effort of time are cut down to a minimum level - especially under consideration of the recommended maximal magnet cool-down rate of 10 K per 30 min.

5.1.2 Cryostat

It is necessary to operate the entire experiment inside a liquid helium-4 (He-4) bath at ambient pressure, called *helium I state*. In this state its boiling temperature is 4.222 K. With a standard atomic weight of 4.002602 u helium gas tends to leave the terrestrial homosphere to accumulate in the heterosphere. At the current price of 6,84 €/m³[1] for helium gas it is most convenient to collect and recycle vaporised helium that leaves the cryostat. To prevent fire hazard due to oxide condensation along the exhaust pipes, exhaust helium is first led into a heat exchanger and afterwards passed into the exhaust gas recirculation, where it is finally purified and liquefied in the on-site 2 kW refrigerator at the ITEP. The actual price for the internal liquefaction is 3,52 €/l[2] liquid helium. With a density of 125 g/l at boiling point[3] [Wik16] and 178.6 g/m³ at standard temperature and pressure[4] [Wik16], the volume from liquid to gas phase increases by a factor of 700.

1 Linde pricing (March 2016), verbal communication with S. Holzmann, ITEP
2 KIT pricing (March 2016), verbal communication with S. Holzmann, ITEP
3 4.2 K, liquid phase
4 0 °C, 1013 hPA, gas phase

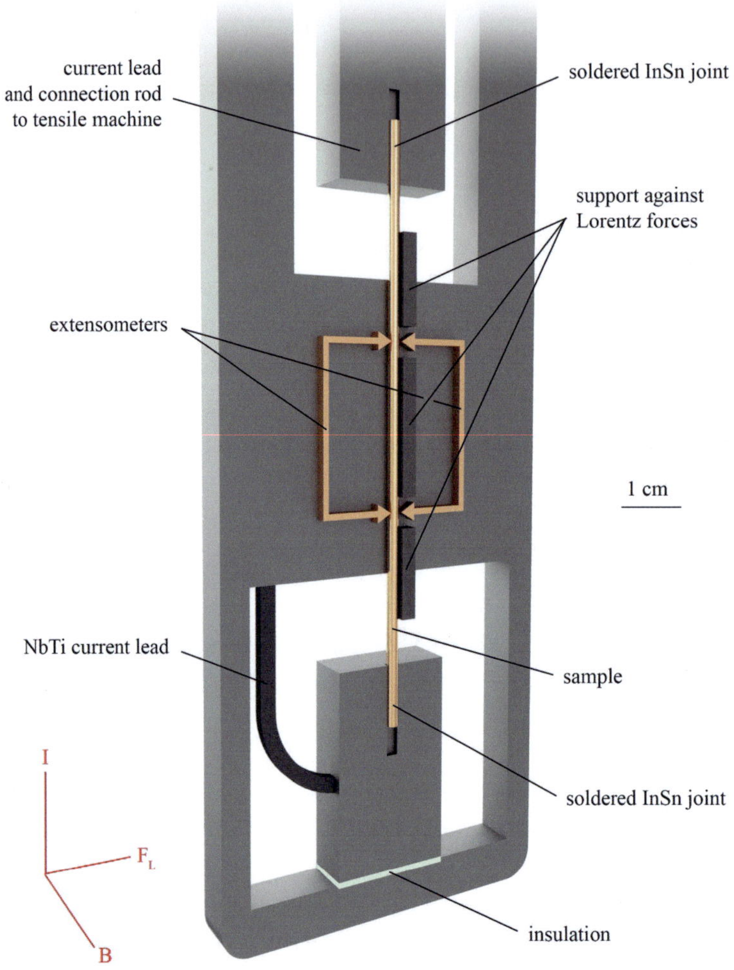

Figure 5.2: Schematic of the FBI sample holder. Picture from [Sch15].

In order to prevent the liquid helium from vaporising rapidly inside the helium chamber, it needs to be shielded against thermic radiation and thermal conduction. This is provided by a liquid nitrogen shield and an isolation vacuum of 10^{-4} Pa. In case of spontaneous helium volatilisation, as it is experienced during a magnet quench, the helium exhaust gas recirculation can be bypassed and the helium is stored inside a balloon storage with a gas volume of 2,000 m^3. In case of excess pressure inside the helium chamber, a pressure relieve valve is triggered at 0.17 MPa. Within this work, the existing cryostat was connected to the nitrogen and helium supply and re-integrated into the exhaust gas recirculation.

5.1.3 Superconducting magnet

The core of the single wire FBI facility is the 14 T superconducting split-coil magnet. It consists of outer NbTi and inner Nb_3Sn conductor windings. Within this work the magnet was removed from the FBI-T facility and integrated into the FBI facility to increase the volume for the sample holder. The split-coil design provides a 70 mm wide and 20 mm thick vertical sample gap, aligned perpendicularly to the magnetic field direction. Due to the high inductance of 154 H the peak magnetic field can be achieved with currents far below 150 A.

Although the magnet is not a solenoid, but a split-coil model, its magnetic field is very homogeneous. Its homogeneity along the centre line is demonstrated in figure 5.3 and its radial field homogeneity is visualized in figure 5.4.

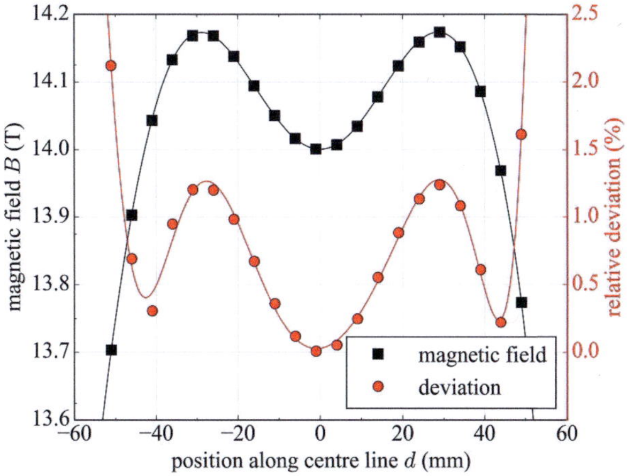

Figure 5.3: Magnetic field strength and relative deviation in the axial direction with polynomial fit function. The zero point represents the centre of the magnet. Data from [Oxf].

In the axial direction the field deviation is 2.1% along the complete measured distance of 100 mm. Since the actual sample thickness is around 2 mm, the accounting deviation is less than 0.2% and can be neglected. In the radial direction the relative deviation is of higher value. Considering a distance of 5 cm between the voltage taps, the radial field deviation is 2%. Although it is a tolerable deviation, it cannot be ignored and needs to be considered in the data analysis.

During the integration of the magnet, one of the most important steps was the installation of the quench resistors (figure 5.1). Although these resistors are not mandatory for regular operation, they become important during a magnet quench, when the superconducting NbTi and Nb_3Sn wires shift from Shubnikov phase into normal phase. The dissipated heat due to ohmic heating can easily damage or destroy the magnet components. Therefore, during a quench, the affected magnet section is short-circuited such as the current dissipates into heat at the quench resistors outside the magnet. An active trigger to open the short circuit is not required. During normal operation the high inherent resistance of the quench resistors directs the current along the magnet windings. In case of a quench the resistance along the magnet winding exceeds the resistance of the quench resistors and redirects the current.

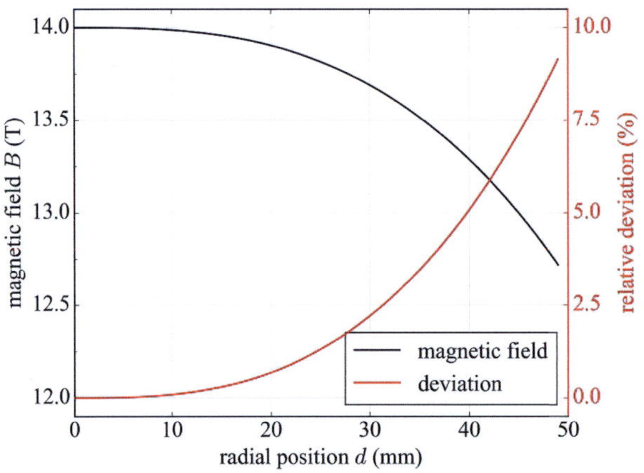

Figure 5.4: Magnetic field strength and relative deviation in the radial direction. The zero point represents the centre of the magnet. Fit function from [Oxf].

5.1.4 Tensile machine, power supply and sensor system

In order to apply tensile strain on the measured sample, a tensile machine with a maximum tensile load of 10 kN can be connected to the fixture port. To achieve precise results the machine position should only be adjusted under consideration of the distance measured by the extensometers. Internal control units are not sufficiently precise. Tensile loads can be measured with the load cell on top of the sample holder.

A 300 A or alternatively a 1.5 kA power supply provided the sample current. Both power supplies are controlled by the FBI-T measurement software.

The sensor periphery includes two Cernox^{TM} sensors to measure the magnet's temperature during cool down and to indicate a sufficient helium level to operate the magnet. Only a measured temperature of 4.2 K guarantees full submergence of the magnet inside the helium bath. All sensors, including temperature sensors, voltage taps and extensometers are read out by the measurement instruments from the FBI-T facility which will be described in detail in section 5.2.4.

5.2 Sub-size cable FBI-T measurement facility

A schematic visualisation of the complete FBI-T facility can be seen in figure 5.5. Originally it was designed in 1988 and constructed to measure LTS sub-size cables for ITER magnet cables. Six years ago the analogue data acquisition system was replaced by digital measurement equipment and a software based control unit.

Within this work a further upgrade was executed, aiming for optimisation to measure HTS strands as they could be used inside a DEMO full-size cable for a TF coil design. The former 14 T superconducting magnet was replaced by a new 12 T split-coil magnet with more than double the sample gap volume. This was required to adept to the increased diameter of new HTS cable designs and to optimise the temperature variable insert (section 5.2.5) for measurements at temperatures of 77 K. After the magnet upgrade, the bottom structure including cross-head and both sample grips were modified, the G10 isolation at the helium outlet was replaced to reduce helium leakage and the overall sensor unit was adapted to the modifications.

With an overall length of 116 cm and a maximal recommended sample diameter of 2 cm, the samples come with inherent stability and only need to be mechanically reinforced against Lorentz forces. This is mostly done by a temperature variable insert (section 5.2.5) or a simple G10 former.

Cuboid copper terminations at both ends of each sample allow for a mechanical fixture by clamping the sample terminations into the designated FBI-T copper terminals. For tensile tests, the sample is further gripped by a stainless steel structure above the bottom and below the top copper termination. The upper sample grip is connected to the tensile machine and can be moved in the vertical direction together with the corresponding copper terminal.

Longitudinal elongations are measured by two CryoMak copper-beryllium extensometers which are located above the magnet and measure the elongation over 48 cm, which corresponds to 86% of the tensile strain experiencing cable length.

Detailed sample specifications and dimensional requirements for FBI-T samples can be found in annex A.2.

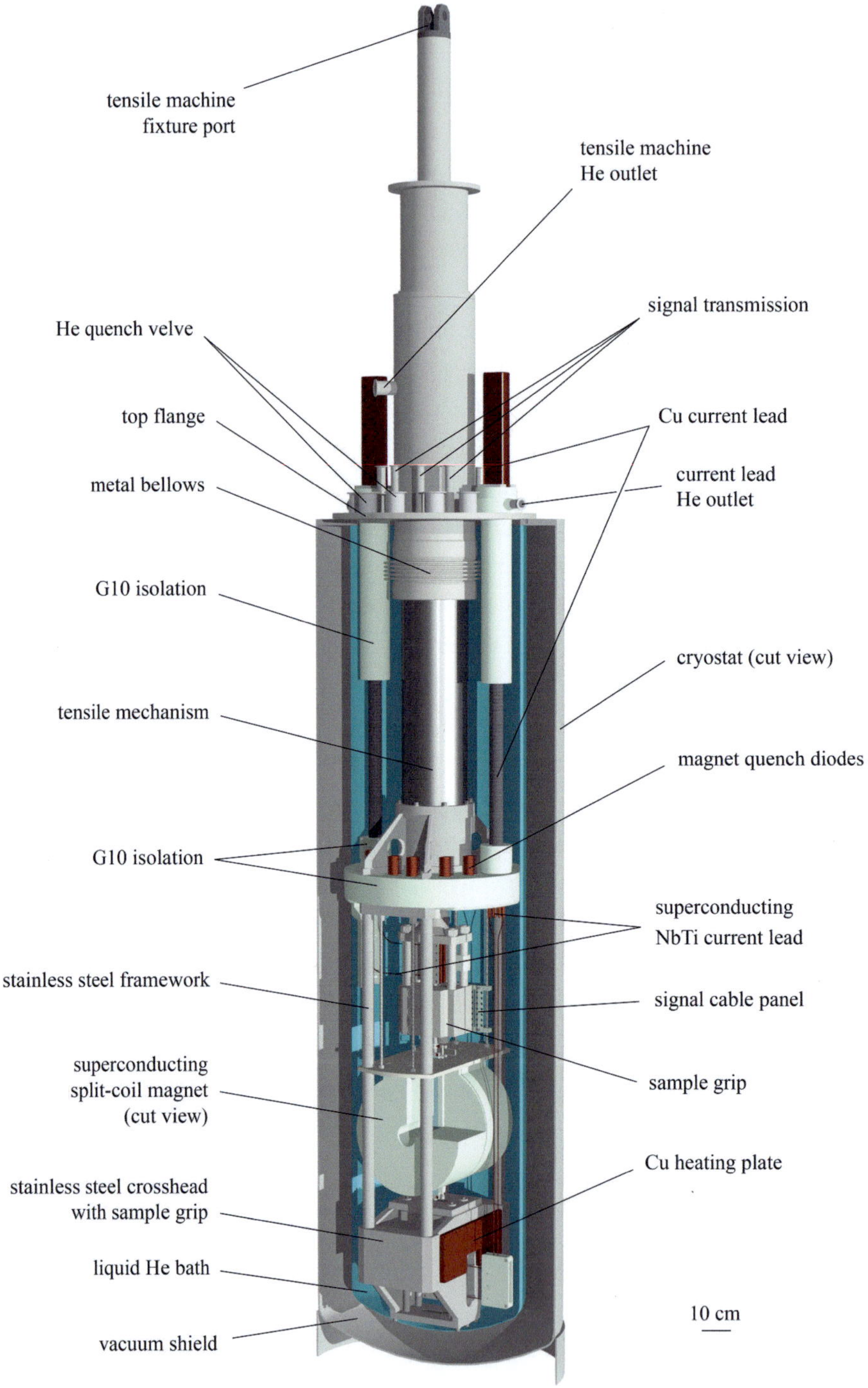

Figure 5.5: Cutaway view of the FBI-T measurement facility.

5.2.1 Sample holder

The mounting principle of an FBI-T sample is demonstrated in figure 5.6.

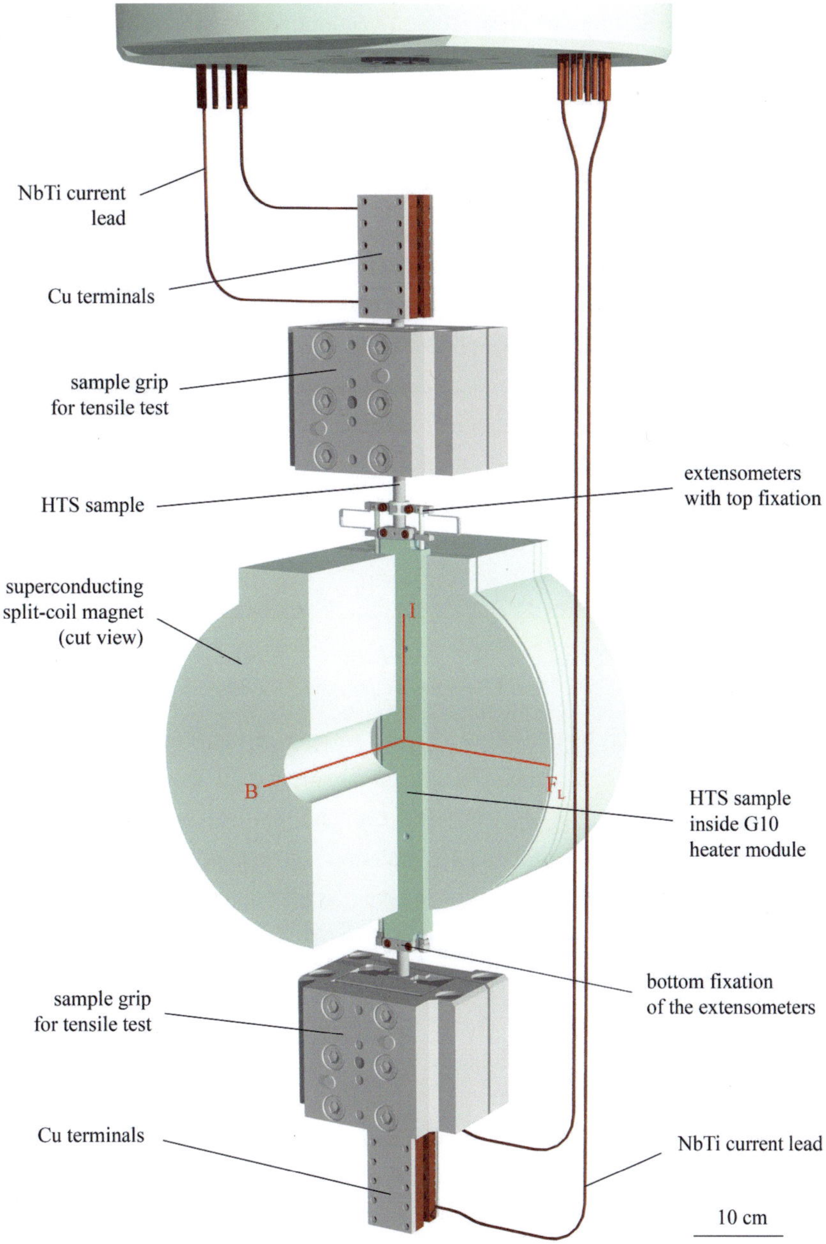

Figure 5.6: Schematic of the FBI-T sample section.

5.2.2 Cryostat

For the same reasons as already mentioned in section 5.1.2, the cryostat is separated into two chambers: the innermost chamber, which contains the liquid helium, and the surrounding vacuum chamber, which contains an insulation vacuum of 10^{-4} Pa. The FBI-T cryostat is moveable. During maintenance it is stored underground. After the proper sample installation it is lifted with a crane and sealed at the top flange (figure 5.5).

The FBI-T cryostat is connected to the helium exhaust gas recirculation, with the additional feature, that the helium flow rate, passing the sample current leads and the magnet current leads, can be adjusted individually. The procedure during spontaneous helium volatilisation remains the same. Therefore only one of both facilities can be operated at the same time. Otherwise, the 2,000 m^3 volume of the balloon storage would be insufficient.

After termination of the experimental phase, the superconducting magnet needs to be heated up to room temperature before opening the cryostat. Otherwise air would condensate inside the magnet and oxidise the magnet windings. In order to heat the magnet to room temperature within two days, two copper heat plates, containing heating elements with a total heat output of 500 W, can be activated (figure 5.5).

5.2.3 Superconducting magnet

The magnet of the FBI-T facility is a 12 T superconducting split-coil magnet. It replaces the former 14 T magnet, which is now installed in the smaller FBI facility. The decrease in peak magnetic field was satisfied by a highly increased sample gap of 80x40 mm^2. Although the large sample gap is an engineering challenge, the magnetic field shows a solid homogeneity in the axial and the radial direction. Field function and relative deviation are visualized in figure 5.7 respectively figure 5.8 with the zero position representing the magnet centre.

The magnetic field in the axial direction, parallel to the field lines, is very homogeneous. For a maximum sample width of 20 mm the relative field deviation is only 0.1% and can be neglected. In the radial direction the field homogeneity is of lower value. Since the sample length of 116 cm exceeds the magnetic high-field zone by far, the sample voltage should be measured within a distance of 10 cm around the high-field zone to reduce the error of measurement. Then again, the limiting factor of the critical current inside the superconducting sample is only the local peak magnetic field. Under the assumption that measured HTS samples do not suffer from local damage, and magnetic effects on the complete sample twist pitch or transposition length are not under investigation, the radial field homogeneity plays a minor role. For samples with twist pitches and transposition lengths that lie within the high-field zone the field deviation has to be considered individually.

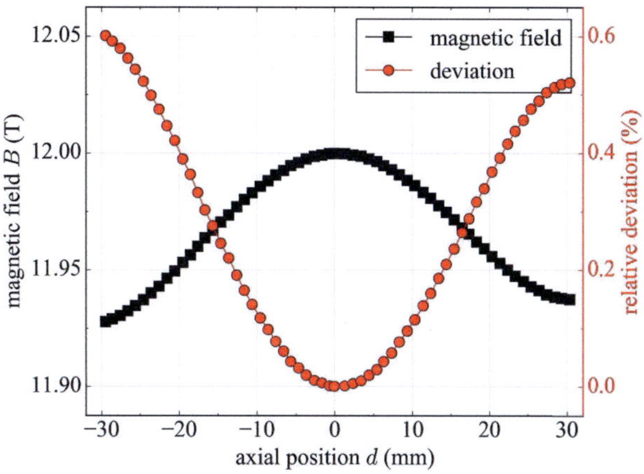

Figure 5.7: Magnetic field strength and relative deviation in the axial direction. The zero point represents the centre of the magnet. Data from [Oxf].

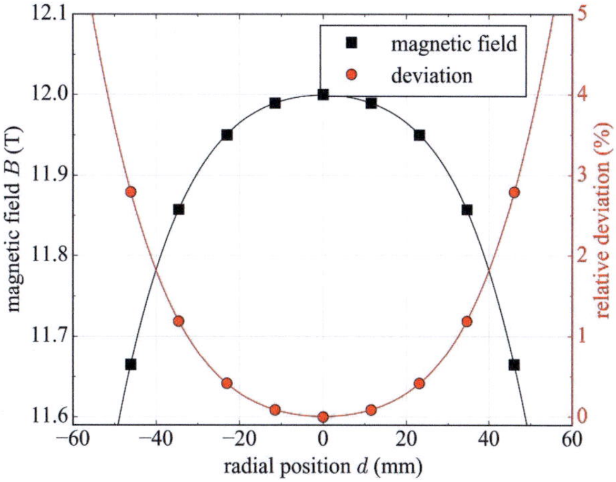

Figure 5.8: Magnetic field strength and relative deviation in the radial direction with polynomial fit function. The zero point represents the centre of the magnet. Data from [Oxf].

5.2.4 Tensile machine, power supply and sensor system

Tensile strain in the longitudinal direction is applied by an electro-mechanic 100 kN tensile machine which is directly connected to the upper sample grip. The tensile machine is adjusted under consideration of the distance measured with the extensometers. Internal control units are not sufficiently precise.

The sample current is provided by a 10 kA DC power supply. To keep the superposition of alternating currents to a minimum level, the DC power supply consists of water-cooled transistor banks. The power supply is controlled by the FBI-T measurement software. In case of a sample quench, the power supply interlock can be triggered by individually adjustable software and hardware quench criteria.

The sensor periphery of the FBI-T facility contains various temperature sensors for the sample, the superconducting magnet, the current leads and the magnet heaters. All of them are read-out by two Lakeshore 218 temperature monitors with maximal 8 sensors each.

Up to four sample voltage taps can be measured with two Agilent/HP 34420A nano-voltmeters. Any further voltage tap respectively the quench voltage is amplified by a Femto DLPVA-100-F-D voltage amplifier and measured by a NI USB 6992 AD converter. Voltages of higher value such as the voltage drop along the copper current leads is directly measured via AD converter.

All collected data is processed in a compiled stand-alone Labview software, which was originally developed during a former PhD thesis [Bar13].

5.2.5 G10 heater module

One of the most important parameters during the critical current measurement is the sample temperature. With increasing temperature the current carrying capability of HTS samples drop significantly. In order to measure the temperature dependant critical current function $I_c(B,T)$, it is necessary to measure and to control the sample temperature. Therefore, a G10 heater module was developed to reach temperatures beyond 77 K. The schematic exploded-view drawing can be seen in figure 5.9.

The core of the heater module consists of two polyimide thermofoil™ heaters. Their intrinsic resistivity is only slightly reduced under cryogenic conditions and they come with a dielectric strength of 1000 V_{RMS} [Min]. With both heaters operating in parallel, the available heat output is 200 W. Both polyimide heaters are fixed on copper plates which conduct the generated heat to the sample, where it is measured by five temperature sensors. Three Cernox™ SD sensors measure the sample temperature at the centre point. They can be operated in static magnetic fields. Furthermore, two temperature calibrated coal resistors TVO are attached at between both ends of the copper contact plates(figure 5.9). They are used to measure the temperature distribution at zero field and help understand the temperature distribution along the HTS sample. The bottom part of the G10 heater module is stabilising the sample against Lorentz forces. Any other G10 component serves for thermal isolation against the liquid helium bath.

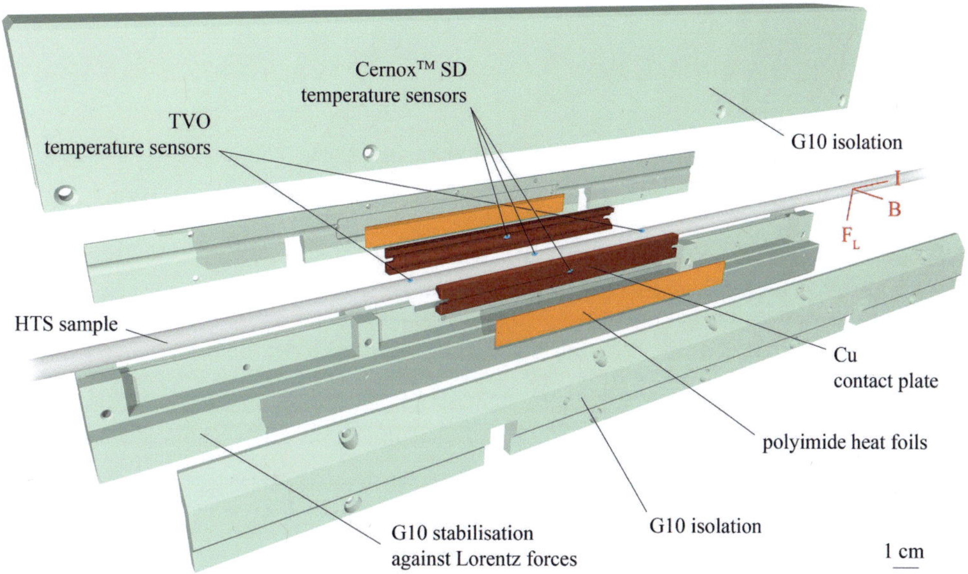

Figure 5.9: Exploded-view drawing of the G10 heater module.

Experience has shown that the measured temperature is highly dependant on the sensor position. While both Cernox™ temperature sensors inside the centre cavities of the copper contact plates measure the same temperatures, the other three sensors measure individual values. In order to understand the temperature distribution inside the samples, the thermal conduction was simulated by FEM. The results will be discussed in chapter 6.

5.3 Uncertainty in measurement

The consideration of statistical and systematic errors during data analysis is very important. In order to stick to international standards, all standard uncertainties for experimentally acquired data are calculated on basis of the Guide to the Expression of Uncertainty in Measurement (GUM) of the Joint Committee for Guides in Metrology [JCG08]. Although the error respectively the uncertainty calculation of the GUM is exactly the same as it is ordinarily applied in physics, this thesis will stick to the precise GUM denominations for reasons of uniformity.

The measurand Y, which in this work is often the critical current I_c of superconductor samples, is defined by the function

$$Y = f(X_1, X_2, ..., X_N)$$

with N measured quantities $X_1, X_2, ..., X_N$.

In case of the critical current I_c it is most likely

$$I_c = f(B, T, \varepsilon)$$

with the magnetic field B, the temperature T and the longitudinal strain ε.

In the following, any estimations of Y and X shall be denoted as y respectively x.

Random error - type A standard uncertainty:

Any scientific experiment is based on the premise to increase its scientific plausibility by reproducing the experimental outcome. However, although the boundary conditions for each repetition remains the same, the results for all independent observations are not necessarily equal and vary within a statistical spectrum.

In order to get to a single and plausible result, it is convenient to calculate the arithmetic mean

$$\bar{X}_i = \frac{1}{n} \sum_{k=1}^{n} X_{i,k}$$

of the measured quantity X_i from n independent observations.

Since the arithmetic mean does not contain any information about the variance of the observed data, it is necessary to calculate the corrected sample standard deviation respectively experimental standard deviation for each measured quantity

$$s(X_{i,k}) = \sqrt{\frac{1}{n-1} \sum_{k=1}^{n} (X_{i,k} - \bar{X}_i)^2},$$

which provides information on the dispersion about the arithmetic mean considering the Bessel's correction. To estimate the standard deviation of the error in the arithmetic mean relating to the expected value, the standard error of the mean (SEM)

$$s(\bar{X}_i) = \frac{s(X_{i,k})}{\sqrt{n}}$$

needs to be considered.

Disregarding any systematic error, the standard uncertainty

$$u(x_i) = s(\bar{X}_i)$$

is equal to the standard error of the mean.

Systematic error - type B standard uncertainty

Beside the random error, the systematic error of measurement also affects the standard uncertainty. In contrary to the random error, the systematic error is predictable, correlates to the measured quantity and can be mostly reduced by optimizing the measurement conditions. Reasons for systematic errors are the method of observation, the imperfection of the measurement instruments and the interference of the observation with the system to be measured. In the majority of cases the manufacturers of measurement instruments and laboratory equipment provide the systematic error. Since the standard uncertainty $u(x_i)$ is defined on the one sigma standard deviation level, which corresponds to a 68% coverage regarding a normal distribution, the systematic error needs to be adapted accordingly.

Combined standard uncertainty

The arithmetic mean of a measurand Y estimated from the measured quantities $X_1, X_2, ..., X_N$ does not necessarily follow a linear function. Therefore, in the following work, the arithmetic mean of the measurand is estimated by

$$y = \frac{1}{n}\sum_{k=1}^{n} Y_k = \frac{1}{n}\sum_{k=1}^{n} f(X_{1,k}, X_{2,k}, ..., X_{N,k}).$$

In order to estimate the combined standard uncertainty $u_c(y)$ of the measurand, it is necessary to propagate type A and type B standard uncertainties of the independent measured quantities by

$$u_c(y) = \sqrt{\sum_{i=1}^{N}\left(\frac{\partial f}{\partial x_i}\cdot u(x_i)\right)^2} = \sqrt{\left(\frac{\partial y_1}{\partial x_1}\cdot u(x_1)\right)^2 + \left(\frac{\partial y_2}{\partial x_2}\cdot u(x_2)\right)^2 + ... + \left(\frac{\partial y_N}{\partial x_N}\cdot u(x_N)\right)^2}.$$

This propagation also needs to be considered for the additive combination of type A $u(x_i)$ and type B $u(X_i)$ standard uncertainties. Under the premise that both, type A and type B uncertainties are based on the one sigma standard deviation level, they are to be combined to the standard uncertainty $u_c(y)$ by following the above-mentioned formulation.

Applied standard uncertainty

To estimate a realistic type B standard uncertainty $u(X_i)$ for the measurement equipment, following accuracies were considered.

- The uncertainty for the Agilent/HP 34420A nano-voltmeters scales with the reading and the range. At temperatures of $23 \pm 5\,°C$ for a maximal interval of 1 mV the accuracy is 0.0050% of the reading plus 0.002% of the range [Agi03]. Since the measured voltages to measure the critical current never exceeds 100 μV, the accuracies due to reading and range are considered with a conservative accuracy of 25 nV.
- For the Lakeshore 218 temperature monitors in combination with Cernox^{TM} 1050 temperature sensors, the relative accuracy is depending on the sample temperature and the magnetic field and is given in table 5.1.
- The accuracy of the current provided by the 10 kA power supply was measured. The error is estimated 0.5% of the actual current. This is a very conservative estimation although it already includes the accuracy of the NI USB 6229 AD converter with 112 μV in the nominal range of 0.2 V [Nat14].
- For the magnetic field B, the accuracy is 0.5% and was provided by Oxford instruments [Oxf16].
- Uncertainties of measured distances to calculate the critical electrical field E_c are considered with ±0.1 mm.
- The accuracy of the load cell used during pressure tests was adapted from the machine it was calibrated with. The data was provided by Deutscher Kalibrierdienst. The linear data fit results in the uncertainty function $0.0017 \cdot x + 0.0013$, where x is the applied load in kN.

Table 5.1: Accuracy of Cernox^{TM} 1050 temperature sensors. Data from [Lak].

	magnetic field B (T)		
temperature T (K)	2.5 T	8 T	14 T
4.2	0.1%	-0.15%	-0.85%
10	0.04%	-0.4%	-1.1%
20	0.04%	0.02%	-0.16%
30	0.01%	0.04%	0.06%
77	0.002%	0.022%	0.062%

Although all the accuracies are well known, the overall type B standard uncertainty can be simplified. The error propagation for the calculation of the critical electric field E_c is

$$u_c(E_c) = \sqrt{\left(\frac{\partial \frac{U}{d}}{\partial U} \cdot u(U)\right)^2 + \left(\frac{\partial \frac{U}{d}}{\partial d} \cdot u(d)\right)^2}.$$

With increase of the distance d between the voltage taps, the uncertainty decreases. The minimum value of d is 10 cm, which corresponds to the high field zone of the magnet. Under consideration of a one-microvolt-criteria the maximal relative type B standard uncertainty $u(E_c)$ is only 0.0056% and can be neglected.

The error propagation for the calculation of the n-value and the critical current is more complicated. The n-value is calculated on basis of equation 3.2 by the method of least square

$$\sum_{k=1}^{k} (ln(E) - (n \cdot ln(j) + c))^2 \rightarrow min$$

which results in the fit parameters

$$n = \frac{\frac{1}{k}\sum_{k=1}^{k} ln(E_i)ln(j_i) - \frac{1}{k}\sum_{k=1}^{k} ln(E_i) \cdot \frac{1}{k}\sum_{k=1}^{k} ln(j_i)}{\frac{1}{k}\sum_{k=1}^{k} ln(j_i)^2 - \left(\frac{1}{k}\sum_{k=1}^{k} ln(j_i)\right)^2} = \frac{\overline{ln(E)ln(j)} - \overline{ln(j)} \cdot \overline{ln(E)}}{\overline{ln(j)^2} - \left(\overline{ln(j)}\right)^2}$$

and

$$c = \frac{1}{k}\sum_{k=1}^{k} ln(E_i) - \overline{n} \cdot \frac{1}{k}\sum_{k=1}^{k} ln(j_i) = \overline{ln(E)} - \overline{n} \cdot \overline{ln(j)}.$$

The error propagation to calculate the type B standard uncertainty is

$$u(n) = \sqrt{\left(\frac{\partial n}{\partial j} \cdot u(j)\right)^2 + \left(\frac{\partial n}{\partial E} \cdot u(E)\right)^2}$$

and

$$u(c) = \sqrt{\left(\frac{\partial c}{\partial j} \cdot u(j)\right)^2 + \left(\frac{\partial c}{\partial E} \cdot u(E)\right)^2}.$$

To calculate the combined standard uncertainty of the critical current density, these standard uncertainties have to be further propagated by

$$u_c(j_c) = \sqrt{\left(\frac{\partial e^{-\frac{c}{n}}}{\partial c} \cdot u(c)\right)^2 + \left(\frac{\partial e^{-\frac{c}{n}}}{\partial n} \cdot u(n)\right)^2} = \sqrt{\left(-\frac{1}{n} \cdot e^{-\frac{c}{n}} \cdot u(c)\right)^2 + \left(\frac{c}{n^2} \cdot e^{-\frac{c}{n}} \cdot u(n)\right)}.$$

For technical superconductors experience has shown that n-values typically lie between values of 30 to 50 and for the c-parameter accounts $c > -300$. In consequence, due to the error propagation, the combined standard uncertainty $u_c(j_c)$ is smaller than the initial value of the type B standard uncertainty $u(j)$ of the current measurement. Although the error propagation is the mathematically correct approach, for all following measurements, an estimation of an upper error limit is satisfying and used instead. The type B standard uncertainty of the current measurement $u(j) = 0.5\,\%$ is taken as the type B standard uncertainty for the critical current calculation and the calculation of the n-value.

The process to calculate the n-value is not standardised, yet. As mentioned in section 3.1.3, the n-value for measurements in this thesis is calculated from $E = 1\,\mu\text{V/cm}$ to $E = 10\,\mu\text{V/cm}$, resulting in the n_{E1-10}-value. In other scientific publications, n-value calculations often include a smaller interval. In order to estimate the effect of different considered intervals, the n-value given in following measurements include an error estimation which considers deviations due to different n-value calculations. The impact of the n-value calculation due to different intervals from five to ten in natural numbers can be seen in figure 3.1.3 for 36 different technical superconductor samples.

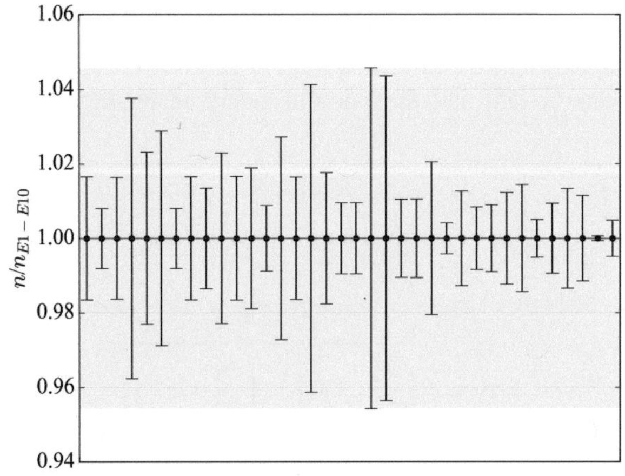

Figure 5.10: n-value calculations for 36 technical superconductor samples. The red line indicates the n_{E1-E10}-value. Error bars represent the average standard uncertainty due to the ten different considered intervals.

Each data point represents a superconductor tape measurement. Error bars indicate the error due to n-value calculations on basis of different intervals taken into account. The red line is the n_{E1-10}-value and the red area represents the peak of the calculated standard uncertainties with regard to the n-value for each tape. As each tape has a slightly different $U-I$-function, it was not possible to find any mathematical relation between the absolute n-value and the deviation due to the considered spread of the electric field during the calculation. Therefore, the maximum estimated error of 4.6% is taken as very conservative error for following n-value measurements. Although this approach does not provide the most precise and narrow n-value error spread for each individual measurement in this thesis, it gives the reader an upper limit of the n-value uncertainty and therefore allows a more realistic comparison of these n-values with measurements from different publications.

5.4 Conclusion

The preparation of both FBI facilities to measure high temperature superconductor samples was successfully executed. While the modernisation of the single tape FBI facility was motivated by the participation in a round robin test, aiming for the standardisation of superconductor measurements under consideration of longitudinally applied strain, the FBI-T facility was upgraded to be able to measure HTS cable strands as they can be used for full-size HTS DEMO cables, including an optimized G10 heater module to achieve temperatures beyond 77 K. All following work and measurements are based on the experimental set-up as it was explicitly described in this chapter.

6 High current HTS cable concepts for DEMO magnet TF coils

High temperature superconductors are a promising alternative to low temperature superconductors for reasons already discussed in chapter 3.5, Due to the high temperature superconductor tape structure, the cable design of a high temperature superconductor cable is completely different to a low temperature superconductor cable design.

Various high temperature superconductor cable designs have been proposed, investigated and optimised within iterative processes during the last few years. The cable concepts to be discussed, analysed and finally evaluated for use in DEMO magnet toroidal field coils are

- the Roebel Assembled Coated Conductor (RACC) cable,
- the Twisted Stacked Tape Cable (TSTC)
- and the Conductor on Round Core (CORC).

6.1 Roebel Assembled Coated Conductors (RACC)

The first high temperature superconductor Roebel Assembled Coated Conductor (RACC) cable was proposed by Siemens, Germany, in 2004, manufactured with 1G $(PbBi)_2Sr_2Ca_2Cu_3O_{10+x}$ HTS tapes [HOLN04]. One year later, the first 2G HTS RACC cable was manufactured at the Karlsruhe Institute of Technology (KIT), Germany [GNK⁺06] and has ever since been developed and optimized at the KIT and at Callaghan Innovation, New Zealand, former Industrial Research Limited (IRL) [BGB⁺16, GFK⁺09, LBB⁺08].

Originally, the RACC cable is based on the Roebel bar, a cable concept to reduce alternating current (AC) losses in copper stator windings of AC generators, which was invented 1912 by Ludwig Roebel [Fra63]. However, in contrary to the Roebel bar it is manufactured of multi-layered *RE*BCO tapes which are punched into a meander shaped form. A schematic of a RACC cable can be seen in figure 6.1.

The geometry of the punching tool depends on the desired transposition length. An increase of the transposition length allows for a higher number of *RE*BCO tapes, while a decrease in the transposition length reduces AC current losses. Since AC current losses inside a toroidal field coil are of secondary importance, the transposition length for all RACC cables manufactured during this thesis was set to 110.5 mm to achieve an assembly of 16 4 mm wide *RE*BCO tapes per cable.

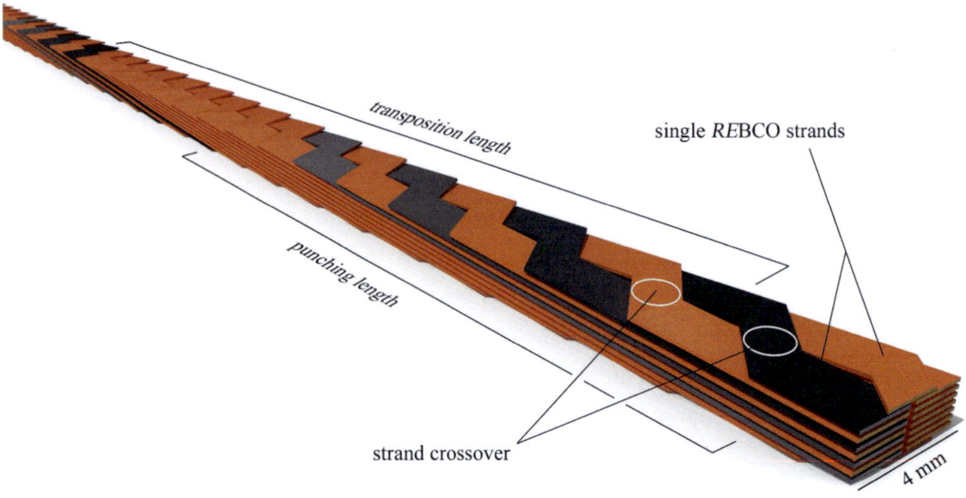

Figure 6.1: Schematic of a 16 strand RACC cable. The punched superconducting tapes were assembled to a Roebel composition. Over the length of two transpositions, each tape is visible once on the top and once on the bottom side. The transposition length is twice the punching length. Coloured strands help to guide the eye.

6.1.1 Terminations

Terminations of a high temperature superconductor cable plays an important role to guarantee a homogeneous current distribution within all tape strands. The statistical analysis on HTS face-to-face lap joints was already discussed and evaluated in chapter 4 and is transferable to face-to-copper joints.

In order to investigate terminations for high temperature superconductor cables the RACC cable is the optimum candidate. Compared to a Twisted Stacked Tape Cable (TSTC), the RACC requires a higher current input homogeneity with most equal joint resistances for all single tapes. The reason for such strict requirement is the bad current sharing within the tapes, visualized in figure 6.2. The vertical gap between the left and the right half of the RACC cable prevents a horizontal current sharing between the tapes. According to the tape orientation of the desired face-to-copper lap joint, the current has to enter from the bottom side to avoid the high ohmic substrate layer on the top.

In case the current is not distributed into all RACC strands homogeneously, the current carrying capability of the cable is highly decreased. This behaviour was described in detail at [Bar13].

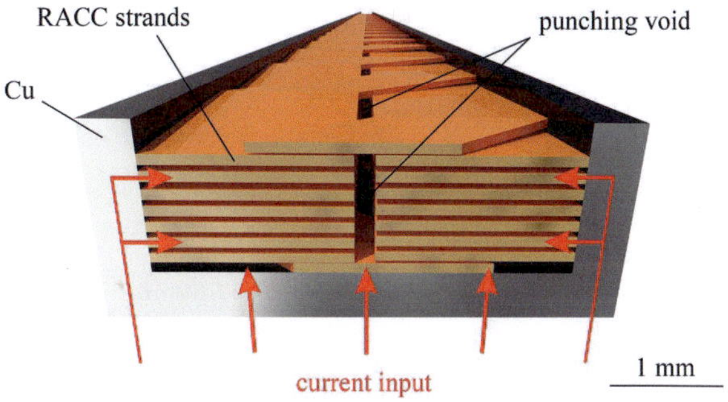

Figure 6.2: Schematic of a RACC cable inside a copper termination. The HTS tapes are oriented with the *RE*BCO layer to the bottom side, to avoid the intrinsic resistance of the high ohmic stainless steel substrate. Voids between the tapes and the copper former are filled with soft solder.

State-of-the-art

One method to avoid inhomogeneous current distribution inside a RACC cable is to splice the cable and solder each tape to the copper termination individually (figure 6.3). This is a very time-consuming process, since each tape needs to be shortened individually. Furthermore, during the solder process each tape is fixed by a stainless steel plate to prevent the tapes from moving while cooling down the termination from solder temperature to room temperature. This additional stainless steel plates increase the termination in width and height. Therefore, this method is recommended for terminations, where space plays a minor role and an optimum low ohmic contact resistance is the decisive factor. Current leads for measurement systems that are cooled by industrial cryo-coolers, which normally are limited to a cooling power of less than 10 W at 4.2 K, are exemplary applications.

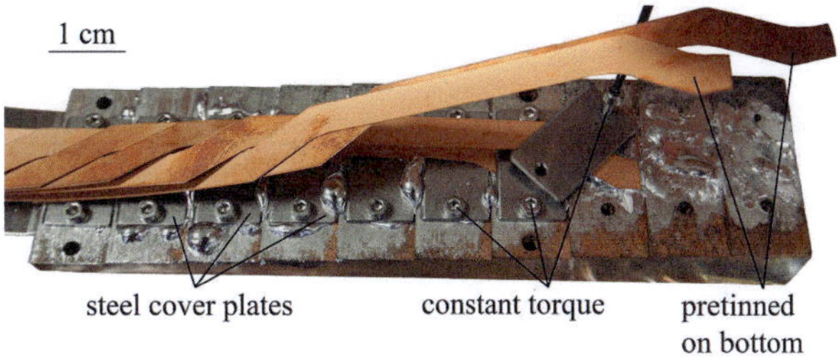

Figure 6.3: Solder joints for a 12 mm wide RACC. Each tape is soldered individually and afterwards fixed with stainless steel plates. Photo from [Bar13]

Optimization of the termination geometry

Since space is the limiting criteria for a fusion cable, it is important to keep the termination to a minimum size. This requires a reduction of the termination width by removing the threads and a limitation of the termination length to the transposition length of the RACC cable. The transposition length is the minimum required length to ensure that each HTS tape is in direct contact with the copper termination. The design of the termination can be seen in figure 6.4. It is separated into two copper parts. As the bottom half-shell (groove) contains the RACC inside the groove and directs the current into the HTS tapes, the top part (tongue) serves as pressure stamp during the solder process. Naturally, for the tongue thickness t_t accounts

$$t_t > d_g - t_{RACC},$$

with the groove depth d_g and the cable thickness t_{RACC}, to prevent the copper structure from absorbing the applied pressure.

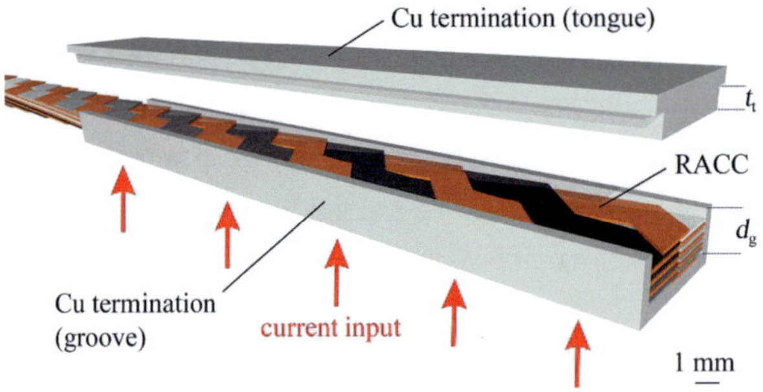

Figure 6.4: Copper termination for a 16 strand HTS RACC. The cable is soldered into the bottom copper groove, while the copper tongue serves to provide pressure during the solder phase. Coloured strands serve a better visualisation of the Roebel structure.

Soldering process

For reasons already discussed in subsection 4.2, In50Sn50 is used as solder and colophony, dissolved in isopropyl alcohol with a mass ratio of 1:2, is used as flux. Since the copper and the HTS superconductor tapes had to be prepared separately, the solder phase is subdivided into three steps.

1. To remove the high ohmic copper oxide, the copper termination is put into a bath of citric acid for half an hour. Afterwards, the citric acid is washed off with distilled water and the termination is cleaned with isopropyl alcohol from grease and residues. In order to pre-tin the cleaned copper termination, it is put on a hot plate and heated to a temperature

of 120 °C. Then the copper is covered with colophony and InSn in turns, until the copper is thoroughly pre-tinned with solder. Excess solder is drawn off with a solder iron and colophony is washed off with distilled water. Since the distilled water cools down the termination below the liquidus temperature, the process needs to be repeated several times until no colophony residues are left. This process is executed for both, the bottom and the top half of the termination.

2. Citric acid cannot be used for its low ph value. To clean the coated conductors from grease and residues, they are wiped with isopropyl alcohol and a soft laboratory cleaning wipe. Afterwards, the RACC is heated to 120 °C on the hot plate. It is important to heat only the cable section which will be pre-tinned. All other cable sections must not be in direct contact to the hot plate, to prevent the copper stabilisation of the coated conductors from oxidation. As soon as the RACC reaches the desired temperature it is doused in colophony. The InSn solder should be thoroughly applied with a solder iron, to make sure that the RACC is entirely pre-tinned on the top- and on the bottom side. Finally, the colophony residues are washed off with distilled water.

3. The third and last step is soldering the RACC into the termination. Both, the copper groove and the copper tongue are heated to 120 °C on the hot plate. When the temperature is achieved, the RACC is placed inside the groove and covered with plenty of InSn until no voids are left. Then the copper tongue is used to close the termination and pressure is applied on top of it to decrease the solder thickness. 6 MPa would be the optimum pressure, however, the pressure to be applied is highly depending on the cable design. As we will see in section 6.1, the RACC without additional mechanical stabilisation is highly vulnerable to pressure perpendicular to the cable plane. Therefore a pressure of only 109 kPa was applied.

A finished solder joint with a length of 140 mm can be seen in figure 6.5.

Figure 6.5: Finished termination for a RACC cable. The termination is oversized to allow for screws and threads which connect the termination to the stainless steel reinforcement for an in-field measurement. Inside toroidal field coil winding, the termination is supposed to be few millimetres wider than the cable width.

Although the tape is only 4 mm wide, the copper termination is 15 mm wide. The reason for the excess width are the threads which are needed to connect the terminations to the stainless steel structure which was used to reinforce the RACC against the Lorentz forces during a measurement in the FBI facility.

Measurement and results

After improving the termination and the solder process, the two terminations were applied to a 40 cm long RACC sample to measure the current input homogeneity. Figure 6.6 demonstrates the experimental measurement setup. Each of the ten strands was measured individually with a voltage tap over a length of more than 40 cm, including the terminations. For this purpose, the taps were soldered outside the terminations as it is proposed in [VGS$^+$13]. This method comes with two benefits: voltage taps applied outside the terminations are easier to mount and any tape degradation inside the terminations due to soldering will be discovered.

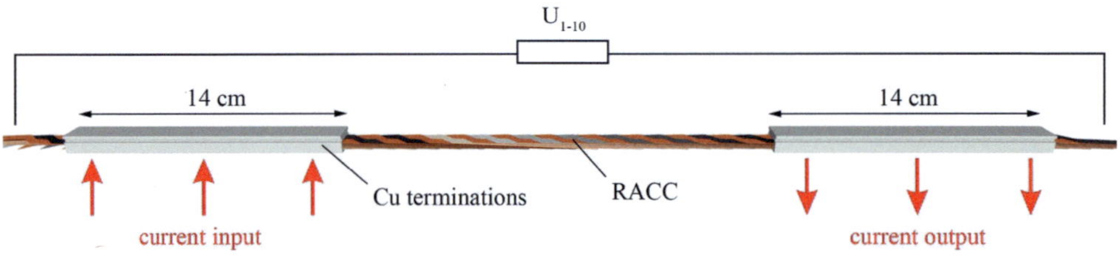

Figure 6.6: Four-terminal sensing of U(I) for a 10 strand 40 cm long RACC. Each tape of the RACC was measured individually. The voltage taps include the terminations. Coloured strands serve a better visualisation of the Roebel structure.

As can be seen in figure 6.7, the current distributes very homogeneously inside the RACC. The functions of the electric field E over the current I for the single tapes are almost congruent. No ohmic contributions were measured at the transition from superconducting to normal state. Unfortunately, voltage tap 7 dropped out during the measurement. However, because the critical current of the whole cable was conform to the critical currents of the single tapes. Solid lines in the graph are power law fits of the measured data, optimized to fit data around the 1 μV/cm critical current criteria. The resulting contact resistance R_{j} is 81 nΩ. This corresponds to a maximal specific contact resistance r_{js} of 453.6 nΩcm^2, when calculating the contact area A_{c} on basis of the 140 mm termination length and the 4 mm cable width, without subtracting the punching voids. Since the punching voids are not subtracted from the contact area, the specific joint resistance is slightly overvalued. However, the punching voids do not affect the overall cable dimensions in any beneficial way. Therefore, it is justifiable not to consider them in the specific joint resistance.

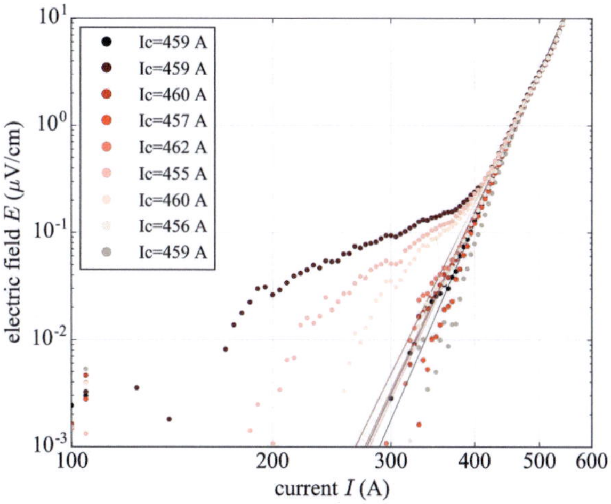

Figure 6.7: Homogeneous current distribution inside a RACC of ten strands. For each single tape, the voltage was measured separately. The current was taken from a shunt and therefore was measured for the overall cable. The power law was used to fit the data points. The electric field was calculated on basis of a 1 μV criteria at the phase transition.

6.1.2 Mechanical reinforcement and pressure resistance

Direct current (DC) applications, such as the DEMO magnet coils, benefit from the RACC cable structure for two reasons. Its tape transposition reduces quench events during magnet charge and therefore reduces magnet training effects. Same accounts for plasma fluctuations which induce current oscillations in the toroidal field coil. Furthermore, the plane-symmetric cable structure of the RACC cable can be aligned parallel to the magnetic field lines. As already discussed in chapter 3.5, the critical current density for parallel aligned *RE*BCO tapes inside an external magnetic field exceeds the critical current for perpendicular alignment by far.

RACC without mechanical reinforcement

The experienced stress inside a magnet winding is hoop-stress and shear-stress. Due to the large magnet diameter of a DEMO toroidal field coil, hoop stress transfers into longitudinal stress on the cable, while shear-stress, as a result of transversal Lorentz forces on the RACC cable, moves the single tapes against each other.

The first measurements of RACC cables in magnetic background fields have proven its high vulnerability to Lorentz forces. Figure 6.8 demonstrates the cable degradation.The cable was enclosed in a G10 former which prevented the cable from bending in direction of the Lorentz force. Still, the critical current value already dropped significantly for low magnetic fields of few tesla. With increase of the magnetic field to 6 T, which corresponds to a Lorentz force of

1.8 kN, the transition to normal phase smoothed out. After increasing the magnetic field to 8 T, the superconducting transition was no longer measurable and the superconducting sample degraded irreversibly. Reason for the cable damage was the tape movement within the RACC sample. The single tapes were pressed against each other at the crossovers, which, as seen in figure 6.9, deformed and damaged the inner radii of the Roebel structure.

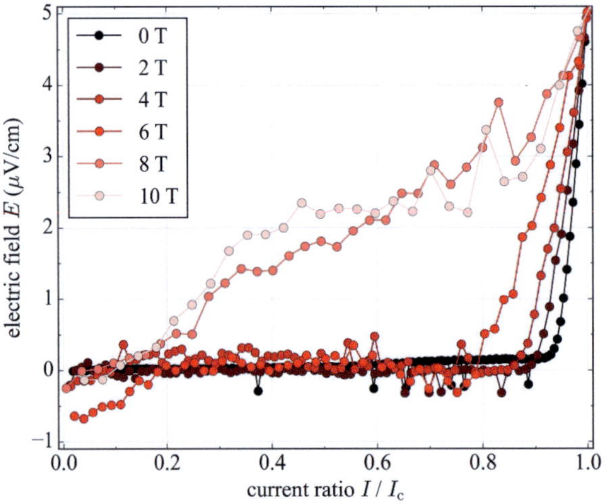

Figure 6.8: Measurement of a ten strand RACC cable at 20 K. The cable was mounted into the FBI facility with the superconductor c-axis parallel to the magnetic field lines. An increase of the magnetic field lead to irreversible cable damage. In order to compare the different functions of the electric field, the critical current was exceptionally estimated with a $5\mu\mathrm{V/cm}$ criteria. Data from [Bar13].

Figure 6.9: Measurement of a ten strand RACC cable at 20 K. The cable was mounted into the FBI facility with the superconductor c-axis parallel to the magnetic field lines. An increase of the magnetic field lead to irreversible cable damage. In order to compare the different functions of the electric field, the critical current was exceptionally estimated with a $5\mu\mathrm{V/cm}$ criteria. Photo from [Bar13].

RACC with mechanical reinforcement

With its very low inherent stability, the RACC cable requires an external reinforcement against shear-stress and tape movement. An earlier solution was the immobilisation of the RACC strands by impregnation [BBW+13]. The composition of Araldite and quartz powder with a mixing ratio of 1:1 has shown a degradation free impregnation at self-field measurements in at 77 K. However, the very low thermal conductivity of Araldite and quartz powder thermally isolate the impregnated superconductor cable from the cooling medium. This is not problematic during normal operation inside a liquid cooling medium bath. However, inside a closed and sealed full-size fusion cable that is operated with helium forced flow, any thermal isolation and reduction of the cross section area of the helium flow highly increases the risk of cable fusing during a quench.

Another approach is to immobilise the cable by mechanical pressure. The only requirement is to induce a large enough friction force to compensate the Lorentz force, induced by the external magnetic field, by compacting the RACC. Compaction does neither reduce the cross-section area of the forced flow nor isolates the superconductors from the cooling medium thermally. For investigation of the RACC compaction, an oversized rectangular stainless steel former, pictured in figure 6.10 was manufactured. The RACC is embedded inside the u-shaped stainless steel former and compressed by the t-shaped stainless steel cover. Stainless steel screws are used to

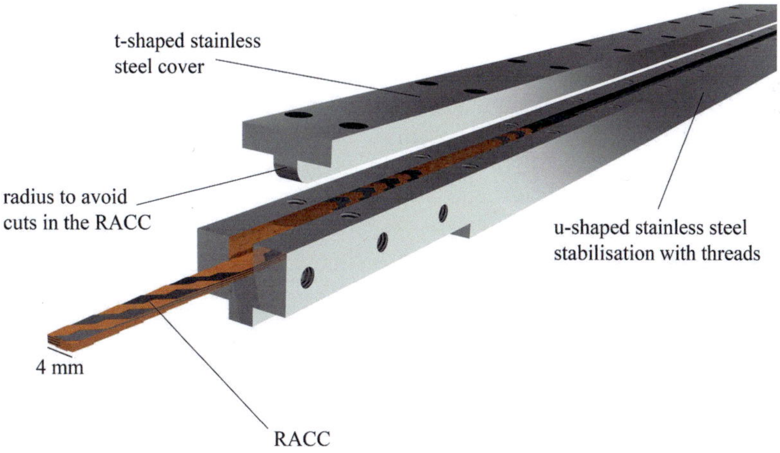

Figure 6.10: Stainless steel compaction for a 4 mm wide RACC cable. The pressure between the t-shaped cover and the u-shaped former is applied by stainless steel screws and induces a friction force of higher value than the expected Lorentz force inside the magnetic background field.

adjust the desired pressure. However, the applied pressure cannot be selected arbitrarily. The friction force induced between the RACC strands has to exceed the Lorentz force. Then again,

*RE*BCO superconductors only withstand a certain amount of transversal pressure. Especially in form of a Roebel composition. To avoid damage during compaction or in-field measurement, the maximum applicable pressure needs to be investigated.

Pressure applicability of a RACC cable

The investigation of the maximum applicable pressure on the RACC cable is very important since pressure is not only experienced during the compaction of the RACC, but also inside a magnetic field. Worst case scenario is the orientation of the RACC cable with the c-axis parallel to the magnetic field lines. In order to understand the dependency of the critical current $I_c(p)$ on the applied pressure, pressure experiments were executed with a 4 mm wide RACC. The pressure distribution depends on the parity of the number of strands. For odd strand numbers, the pressure distribution depends on the angle of the crossovers and is limited to a maximum number of tapes [FBB+15]. With increase of the tape numbers, interferences with high peak pressure occur and destroy the cable. For even strand numbers, the pressure distribution scales linearly with the number of tapes. With increase of the number of tapes, the number of crossovers increases and, as there are no interferences, the pressure is equally distributed [FBB+15].

Sample assembly

Both RACC samples used to investigate the maximum applicable pressure were assembled of ten *RE*BCO strands. They have an overall length of 400 mm which includes both 140 mm long terminations. The minimum termination length is 110.5 mm which corresponds to half the transposition length and guarantees for two direct contacts between each strand and the copper termination.

Sample 1 and sample 2 differ in thickness. While sample 1 consists of 10 Roebel assembled *RE*BCO strands, sample 2 additionally contains 10 0.1 μm thick copper strands. All copper strands were punched into the meander shaped Roebel form and were placed congruently on top of the *RE*BCO strands. By definition the top side of a *RE*BCO tape is the side closest to the *RE*BCO layer. The intention behind the additional copper layer between each *RE*BCO tape is to buffer the pressure sensitive RACC crossovers and to extend the absolute pressure of permanent degradation. A photo of sample 2 can be seen in figure 6.11.

Experimental setup

After successfully assembling both RACC samples, each is mounted into the u-shaped stainless steel former. The pressure during the measurement is applied by two large steel plates which are connected by threaded rods passing the pressure to the 85 mm long t-shaped cover which

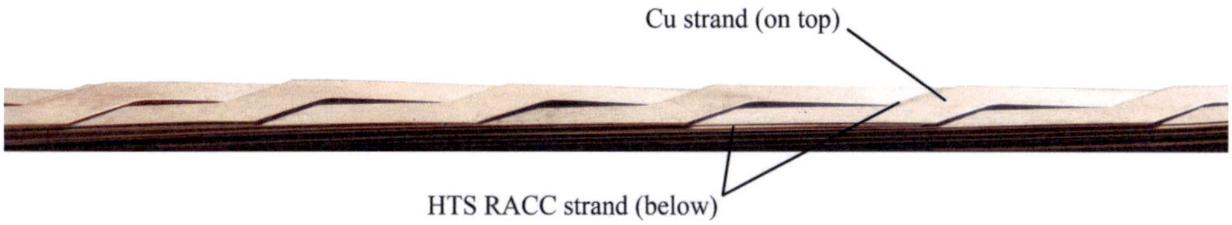

Figure 6.11: RACC sample 2 with 10 REBCO and 10 additional copper strands. Each copper strand covers the top side of one REBCO strand and serves as additional buffer layer against transversal pressure. Slight displacements between the copper and the superconducting strands are visible.

functions as pressure stamp. By smoothly adjusting the torque on each threaded rod, the pressure is increased. A 15 kN load cell with two strain gauges is installed between the steel plates and the t-shaped cover to measure the applied pressure.

During the measurement inside a 77 K liquid nitrogen bath, the pressure is increased from zero. After each increase, the current is ramped to estimate changes in the critical current value. This process is repeated iteratively until the critical current drops significantly. Throughout calibration and measurement, the load cell is fully submerged inside the liquid nitrogen bath to exclude errors due to a temperature gradient along the strain gauges.

The voltage drop of each of the ten strands was measured individually. All voltage taps were installed according to the description in subsection 6.1.1. The corresponding current was measured for the entire sample at the shunt of the power supply.

Pressure estimation

An accurate pressure estimation requires a precise definition of the contact area between the pressure stamp and the RACC samples. In order to compare the results with future measurements and different cable designs, it is reasonable to distinguish three different areas of contact. The different contact areas are recorded in table 6.1. All of them were precisely measured by evaluating high resolution microscope images of the RACC samples.

Table 6.1: Calculated area values based on the contact area definition. All values were precisely measured with high resolution microscope imaging.

	A_{tot} / mm^2	A_{avg} / mm^2	A_{eff} / mm^2
sample 1	340.0	-	95.9
sample 2	340.0	328.6	102.2

Following different contact areas are defined:

- A_{tot} defines the total contact area based on width and length of the pressure stamp. In the experiment the length of the stamp is 85 mm and its width is 4 mm.
- A_{avg} defines the theoretical contact area between the RACC samples and the pressure stamp corrected by punching voids, which reduces the area estimation. The resulting average pressure allows for an comparison of the average pressure between various cable designs.
- A_{eff} defines the contact area, which effectively experiences pressure during the pressure experiment. Due to the meander shaped form of the RACC cable, the pressure is not distributed homogeneously but focused on the cross-sections, which results in the pressure pattern visible in figure 6.12. The effective contact pressure is important to estimate in order to avoid cable damage during the compaction of the RACC cable. The visualisation of the pressure pattern was realized by a 10 MPa pressure measurement film. It was placed on top of the RACC sample and only removed after finishing the pressure experiment. It was also verified, that the increase of the pressure to levels which degraded the RACC samples irreversibly, did not change the pressure pattern.

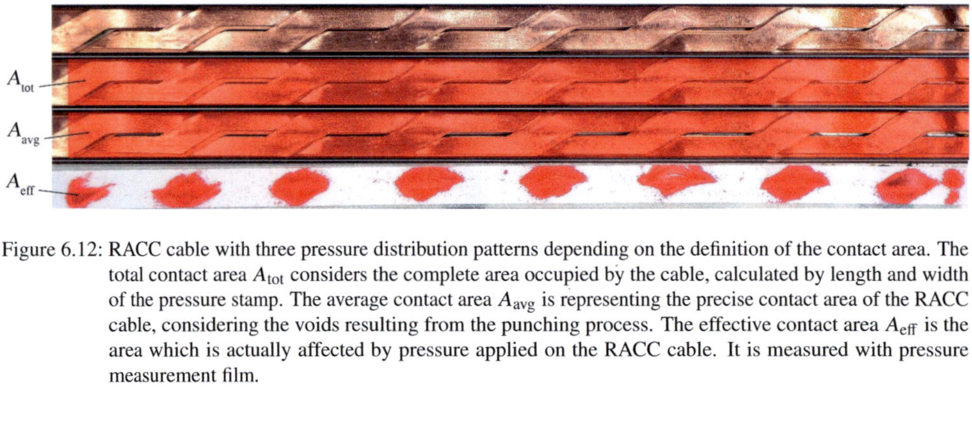

Figure 6.12: RACC cable with three pressure distribution patterns depending on the definition of the contact area. The total contact area A_{tot} considers the complete area occupied by the cable, calculated by length and width of the pressure stamp. The average contact area A_{avg} is representing the precise contact area of the RACC cable, considering the voids resulting from the punching process. The effective contact area A_{eff} is the area which is actually affected by pressure applied on the RACC cable. It is measured with pressure measurement film.

RACC sample without additional copper buffer layers

The very low inherent stability of the RACC and its hight vulnerability against transversal pressure is well known [BBW⁺13, UWB13]. To prove the high vulnerability, sample 1, the 4 mm wide sample without additional copper buffer layers, was measured under transversal pressure. The result of the measurement is visible in figure 6.13. The critical current values are given for both, the average pressure p_{avg} and the total pressure p_{tot}.

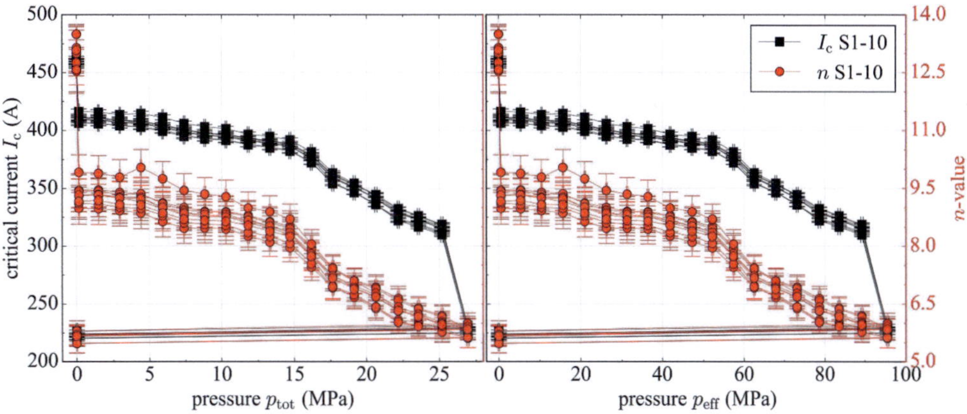

Figure 6.13: Performance of sample 1 at 77 K in liquid nitrogen depending on the total pressure p_{tot} and the effective pressure p_{eff}. The cable immediately degrades by 11% after applying slight transversal pressure. The second significant drop is reached for p_{eff} >90 MPa. The cable damage is irreversible. Connecting lines serve to guide the eye.

The first significant performance drop of 11% occurs immediately after applying slight pressure. Cause for the slight pressure is the 5 kg dead weight of the load cell. This behaviour was reproducible for various 4 mm wide RACC samples. The first applied pressure forces the filigree RACC strands into a fixed position which slightly damages the tapes at the cross-overs. For 12 mm wide tapes with larger proportions between tape thickness and tape width, this effect is not visible. Further increase of the pressure leads to a slightly declining plateau. For effective pressures beyond 60 MPa, the decline increases and drops significantly for effective pressures beyond 90 MPa. The observed degradation is irreversible. The point of irreversibility was not investigated, since the main focus lies on cable compaction, which is a one-time process.

RACC sample with additional copper buffer layers

In order to reduce the vulnerability against transversal pressure, sample 2 contains ten copper buffer strands on top of each superconductor strand. The buffer effect of the additional copper can be seen in figure 6.14.

An initial drop due to the dead weight of the load cell was not measurable. The observed damage at sample 1 was prevented by the additional copper buffer layers. With first pressure increase, the critical current drops about 1.5%. With increase of the effective pressure up to 40 MPa, the critical current remains stable. For values beyond 40 MPa the critical current starts to drop. Only at 120 MPa effective pressure the critical current reaches the 340 A level. Therefore, sample 2 proves a higher durability against transversal pressure compared to sample 1, where the 340 A limit is already reached for 75 MPa effective pressure. The grey bar, denominated as operation point, is the pressure applied during compaction of the RACC. The pressure test resulted in irreversible cable degradation. The ultimate point of irreversibility was not investigated.

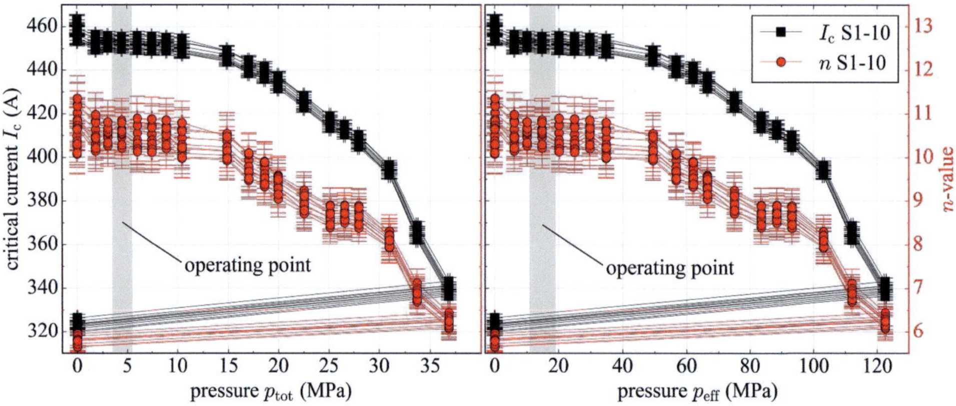

Figure 6.14: Performance of sample 2 at 77 K in liquid nitrogen depending on the total pressure p_{tot} and the effective pressure p_{eff}. The cable damage after the measurement is irreversible. Connecting lines serve to guide the eye.

In order to explicitly compare the results of sample 1, sample 2 with the performance of a single tape, all absolute data was transferred into relative data with the critical current $I_{c0} = I_c(T = 77K, B = sf, p = 0)$ as reference point. The relative data depending on the total and effective pressure are visible in figure 6.15.

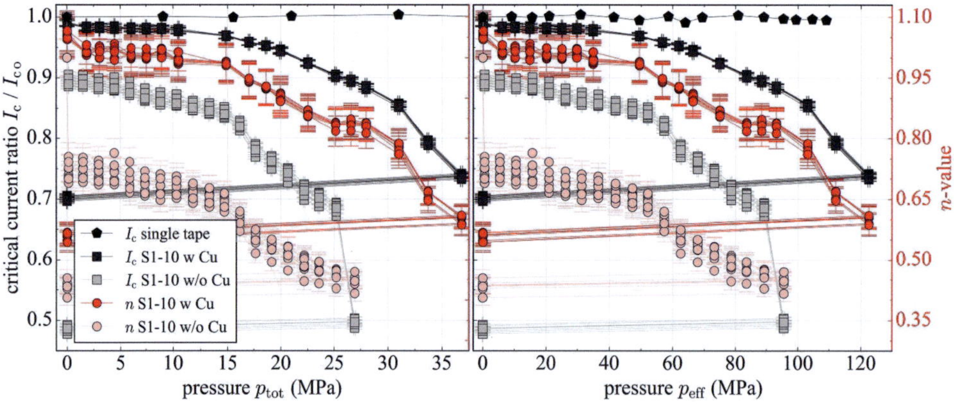

Figure 6.15: Comparison of sample 1, sample 2 and single tape data under transversal pressure at 77 K. The additional copper protects the RACC from degradation due to transversal pressure. Connecting lines serve to guide the eye. Data for the single tape from [TKO⁺07].

By assembling additional copper layers into the RACC, the vulnerability of the RACC could be reduced significantly. The cable damage due to the dead weight of the load cell could be completely prevented and the degradation due to applied transversal pressure extended from 90 MPa to 120 MPa effective pressure. Furthermore, as expected, the single tape exceeds the pressure durability of the RACC. The single tape benefits from its flat, symmetric form which absorbs the pressure homogeneously.

In conclusion, the performance of the RACC under pressure could be improved significantly. Its pressure durability is high enough to withstand the pressure applied inside the stainless steel casing.

6.1.3 Field dependent measurement

After the foregoing pressure investigation, two RACC samples were manufactured, to be measured inside a magnetic background field of 12 T inside the FBI measurement facility.

Sample A was manufactured of ten superconductor strands and ten additional copper buffer strands. To avoid any unforeseen setbacks, the design was exactly the same as for the pressure test. Beside sample A, a second sample, sample B was manufactured. The 4 mm wide design remains the same, but the number of superconductor and copper buffer strands was increased from ten to 16. This is the maximum even number of strands for the pre-defined transposition length of 110.5 mm. Details on both sample parameters can be found in table 6.2.

Table 6.2: Parameters of the FBI RACC samples A and B.

parameter	RACC FBI sample A	RACC FBI sample B
sample length	1160 mm including terminations	
superconductor	340.0328.6	
number of HTS tapes	10	16
number of Cu tapes	10	16
transposition length	110.5 mm	
termination	copper terminations soldered with In50Sn50 soft solder	
mechanical support	compressed inside a stainless steel former	
expected Lorentz force	4.0 kN	6.4 kN
applied torque per screw	0.2 Nm	0.32 Nm
electrical stabilisation	copper tapes on top of the HTS tapes	
pairs of voltage taps	2	4
voltage tap distances	89 cm	56 cm, 89 cm

Stainless steel compaction and sample preparation

In order to correctly calculate the required friction force and therefore the required pressure inside the stainless steel conduit, it is necessary to estimate the Lorentz force on both RACC samples precisely. To do so, the data of the magnetic field of the 12 T magnet, provided by Oxford Technologies, was fitted with the power function

$$f(x) = \frac{1}{a + bx^c}, \tag{6.1}$$

where x represents the coordinate along the sample. The centre of the magnet is defined at $x = 0$.

For the calculation of the Lorentz force, the maximum currents for both samples at 12 T, 4.2 K were estimated with 0.8 kA for sample A and 1.3 kA for sample B. By integrating equation 6.1 over the sample length, the expected Lorentz force over the entire sample length is 3.3 kN for sample A and 5.3 kN for sample B. Considering a very conservative friction factor for copper on stainless steel with 0.23 [Bea13], the required friction forces are 14.4 kN and 23.0 kN respectively. This corresponds to an effective pressure p_{eff} of 16 MPa.

To maintain this pressure during the in-field measurement, the stainless steel cover was closed by several M4 screws. The torque for each screw was calculated by

$$M = F_{\text{pre}} \cdot \left(\frac{d_{\text{p}}}{2} \cdot \tan(\alpha + \beta) + \mu \cdot \frac{d_{\text{h}}}{2} \right),$$

with the preload force F_{pre}, the pitch diameter d_{p}, the thread angle α, the friction angle β, the friction coefficient μ_{f} and the friction diameter d_{h}. The resulting torque for sample A was 0.2 Nm and for sample B 0.32 Nm.

After both samples were completely compacted into their stainless steel conduits and the terminations were manufactured as described in section 6.1.1, the voltage taps were soldered to the RACC samples. Sample A was equipped with two and sample B with four voltage taps. In comparison to sample A, the conduit of sample B was designed with two additional gaps to allow for an additional voltage measurement over a shorter distance of only 56 cm.

FBI sample measurements

After both RACC FBI samples were prepared successfully, they were mounted into the FBI facility with the c-axis of the superconductor parallel to magnetic field lines. It is the worst case scenario with the Lorentz force pushing the edges of the RACC strands against each other.

The measurements of both cables were executed at 4.2 K. Both cable measurements were started with an initial magnetic field of 8 T. This initial value was chosen for two reasons: for lower magnetic background fields, the current density inside the cable is highly increased. In case of a cable quench, the high current density might permanently degrade the cable or even fuse it. The safest approach would be to start the measurement at 12 T background field. The sample current, and therefore the chance of degradation during a quench, is proportionally low. However, the Lorentz force is at its peak. A comparison between measurements before and after reaching the peak magnetic field would not be possible. A change could only be verified by comparing the 77 K measurements in liquid nitrogen before and after the FBI measurements.

Beginning with the 8 T measurement, the magnetic field was increased to 12 T and then cycled between both values in 1 T steps. After any training effect or propagating degradation could be excluded, the magnetic field was finally ramped down to measure the critical current at self-field. The results for the ten strand RACC sample A measurement can be seen in figure 6.16. During the measurement a very slight training effect during the first ramp was observed. However, the effect was negligible and a degradation could be excluded after the subsequent 77 K measurement in liquid nitrogen. Although no degradation was observed, the

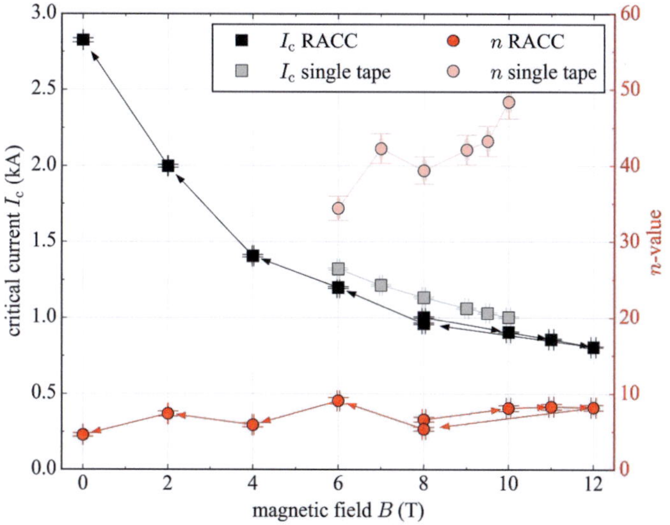

Figure 6.16: Performance of the ten strand RACC sample A during measurement inside the magnetic background field of the FBI facility at 4.2 K. The magnetic field penetrates the sample parallel to the c-axis of the RACC. A slight training during the first ramp to 12 T was observed. The difference between the critical current of the RACC and the up-scaled data of a single RACC tape is approximately 11%. Arrows illustrate the measurement sequence.

RACC cable lagged the performance of the upscaled data of one single punched RACC strand. A hint to the reason is the very low n-value. Both voltage taps used to measure the voltage drop along the RACC were soldered very close to the terminations, to leave enough space for the stainless steel conduit. In case the current does not fully couple into the superconductor over the termination length, its ohmic contribution in the copper layers would be added to the voltage measurement. During the evaluation of the measured data such ohmic contribution is subtracted to estimate the correct critical current value. However, if the current is not fully coupled into the sample, it reduces the steepness of the transition and therefore the n-value. A low n-value makes it very difficult to evaluate the correct ohmic contribution. Therefore it is very probable, that the ohmic contribution was underestimated during the data analysis. This theory was proven by the fact, that the current measurement at 77 K in liquid nitrogen was conform to the self-field simulation. Both will be discussed further in subsection 6.1.4.

To avoid this uncertainty during the second RACC FBI measurement, sample B was equipped with a second pair of voltage taps with a shorter distance of only 54 cm. The FBI measurement of sample B is shown in figure 6.17. The measurement with the voltage taps closer to the magnet high-field zone was very successful. The first critical current data point was acquired for 8 T. Afterwards the magnet was ramped again several times to 12 T. The ramping process can be traced by the development of the n-value, which improved every cycle. This was

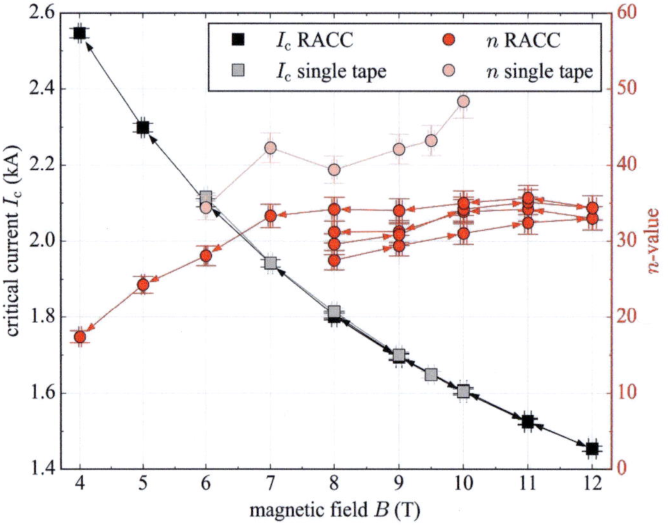

Figure 6.17: Performance of the 16 strand RACC sample B during measurement inside the magnetic background field of the FBI facility at 4.2 K. The magnetic field penetrates the sample parallel to the c-axis of the RACC. No degradation or training effect were observed. The $I_c(B)$-function of the RACC cable is perfectly in accordance with the single RACC tape measurement. Arrows illustrate the measurement sequence.

the only training effect that could be measured. Degradation or damage of the cable was not observed for high magnetic fields. Instead, the $I_c - B$-function of the RACC cable is perfectly congruent to the upscaled, measured single RACC strand data.

Unfortunately, during the 4 T measurement of the RACC sample, the cable quenched.

Although a soft- and hardware quench detection were running in background, the interlock trigger at the power supply could not avoid permanent cable damage. Nonetheless, in a first conclusion it can be said, that the stainless steel conduit and the compression of the RACC conductor could be proven a successful method to reinforce the RACC against Lorentz forces at magnetic fields up to 12 T.

6.1.4 Post investigation

Simulations and measurements at 77 K

Each of the samples was measured at 77 K in liquid nitrogen before compaction and before and after the in-field measurement, to see if any processes harms the cable. The taps to measure the voltage remain the same as during the 4.2 K measurement. To further exclude cable damage during the sophisticated Roebeling process, the expected current for magnetic self-field at 77 K was simulated by finite element method (fem) in COMSOL. Detailed information on the simulation process can be found here [GSZV14]. A visualisation of the expected magnetic field and the current density is shown in figure 6.18. During the simulation, the critical current for RACC sample A was estimated 477 A and for RACC sample B 736 A. The results for the 77 K self-field measurements and the simulations are visually compared in figure 6.19.

For sample A, the measured critical current at self-field without the stainless steel conduit was in accordance with the simulation. After applying the stainless steel conduit, the critical current value of sample A remained constant. After the in-field FBI measurement, sample A showed an improvement of its maximum critical current at 77 K measurement by 20%. This increase is far too high to be explained by a training effect during the in-field-measurement. It is assumed, that the higher critical current is related to changes in the current coupling at the terminations. An improvement in the current coupling reduces the ohmic contribution of the current running along the copper stabilisation. The superconductor transition becomes steeper and the n-value increases, which makes it easier to evaluate the critical current precisely. In this case in favour of the critical current. In order to prevent this uncertainty, as already mentioned, sample B came with different voltage tap distances.

For sample B, the measured critical current at self-field without the stainless steel conduit was also in accordance with the simulation. After compaction inside the conduit, sample B showed a minimal decrease of 4.9%. Although this decrease lies 3% above the expected value of the pressure investigation, this was no reason not to successfully measure the sample in the FBI facility. After the cable was damaged during the 4 T measurement, the critical current at 77 K naturally dropped significantly.

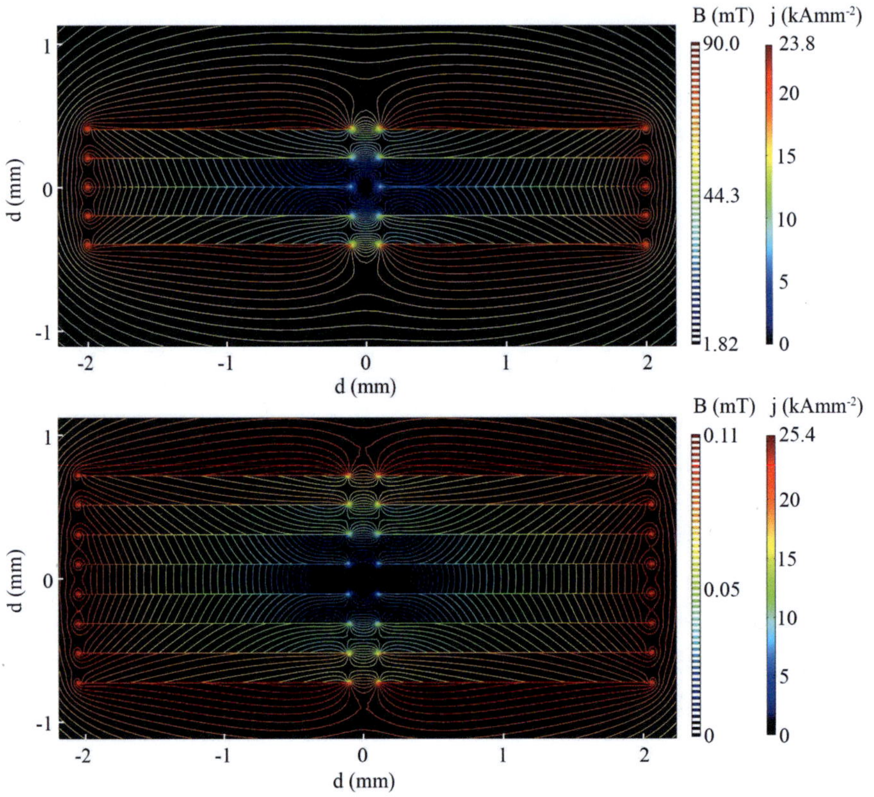

Figure 6.18: Finite element simulation of the magnetic self-field and the current density of the 10 strand RACC sample A and the 16 strand RACC sample B at 77 K. Figure from [BGB+16]

Investigation of delamination effects

Technical *RE*BCO high temperature superconductors are very vulnerable to delamination. The weak spot of the conductor lies between the buffer layers and the *RE*BCO layer. Therefore, already slight delamination most likely damages the superconductor completely and should always be avoided. In case of the RACC, delamination plays a special role. After the punching process the *RE*BCO layer is not fully surrounded by the silver and copper stabilisation layers anymore. The edge on the punching side of the conductor is open. Any external force or load on the open edge easily separates the layers and initiates the delamination process.

After executing all measurements successfully, RACC sample B was removed from its conduit and investigated for delamination effects. The mid-section of the cable, which experienced the magnet's high-field zone, was cut with a laser beam into 16 few centimetre long pieces, which were afterwards pictured with a scanning electron microscope (SEM). Each tape was cut at a different tape section. The SEM visualisation of four random, representative cutting edges can

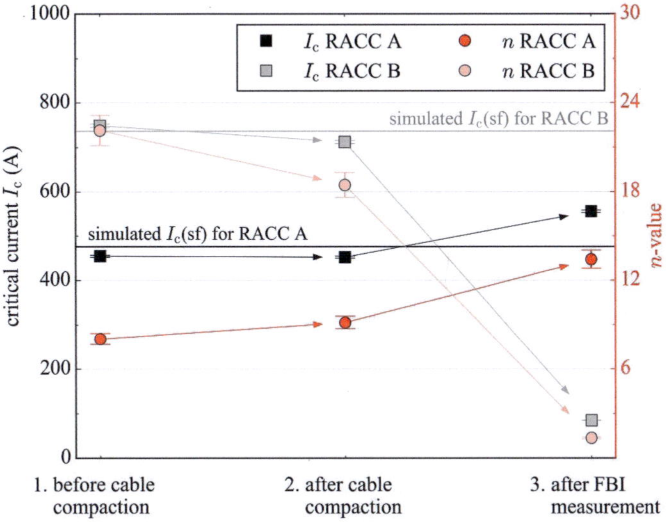

Figure 6.19: Critical current measurements at self-field of both RACC samples before and after the cables were compacted with the stainless steel former and after the FBI measurement. Horizontal lines mark the simulated critical current at self-field. The FEM simulation was executed in COMSOL.

be seen in figure 6.20. The visible tape width varies with the location of the cutting edge. At the cross- over sections the edge is almost twice the size as at the parallel sections. Delamination which might impair the current carrying capability was not detected for any of the investigated tapes. The only visible delamination was observed at edges of very wide cross-over sections close to the outer corner. Due to the over-sized width this does not effect the critical current. In conclusion can be said, that pressure compaction of a RACC cable does not result in tape delamination.

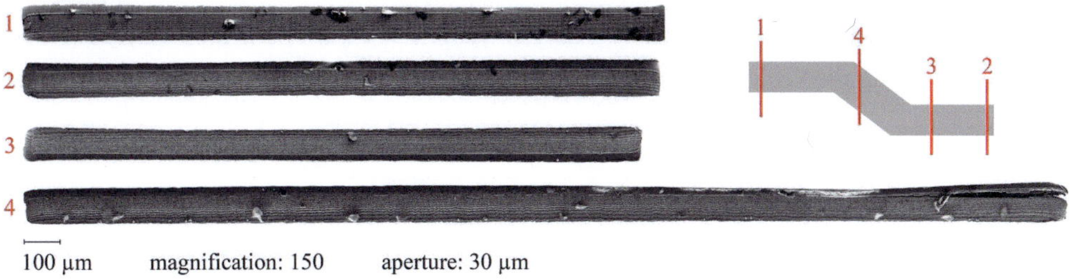

Figure 6.20: Scanning electron microscope (SEM) view of four RACC tapes taken from sample B. All of the 16 strands were cut by laser and checked for delamination effects. Four random represents from different tape sections are illustrated above. Figure from [BGB+16]

6.1.5 Conclusion

The RACC cable is the only transposed cable design among all high temperature superconductor cable designs. It has the lowest alternating current losses and has a flat cable design which allows for a defined orientation of the cable's c-axis inside a magnetic background field. It has been proven, that although the cable is vulnerable to Lorentz forces and transversal pressure, it is possible to reinforce the cable inside a stainless steel conduit with appropriate pre-compression, to successfully prevent damage due to Lorentz forces at magnetic background fields of up to 12 T. The final conclusion on the Roebel Assembled Coated Conductor (RACC) cable is drawn in subsection 6.4, after comparison with the Twisted Stacked Tape Cable (TSTC) and the Conductor on Round Core (CORC).

6.2 Twisted Stacked Tape Cables (TSTC)

The Twisted Stacked Tape Cable (TSTC) is based on the idea to stack single *RE*BCO tapes and twist them along its longitudinal axis. Originally, the idea was proposed at the Massachusetts Institute of Technology (MIT) in 2010 [TMBB10] and was afterwards adapted and modified by various research institutes. Currently, the TSTC design is developed in different variants at *École polytechnique fédérale de Lausanne* (EPFL) in Switzerland, at *Agenzia nazionale per le nuove tecnologie, l'energia e lo sviluppo economico sostenibile* (ENEA) in Italy, at *National Institute for Fusion Science* (NIFS) in Japan and at *Karlsruhe Institute of Technology* (KIT) in Germany. In the following, these different cable designs will be described, investigated and evaluated.

6.2.1 Mechanical stability

Inside a magnetic background field, the longitudinal twisted stacked taped cable experiences Lorentz forces in any radial direction. In case of transversal load parallel to the c-axis, the cable proves to be very robust. The pressure durability of a single tape was already shown in subsection 6.1.2, figure 6.15. However, for loads perpendicular to the c-axis, *RE*BCO tapes are very vulnerable, if not fully stabilised or embedded into a solid former. In-field FBI measurements of a non-soldered and non-compacted TSTC, which was only laid inside a copper tube and fixated by pre-strain in longitudinal direction, had shown complete and irreversible cable degradation. With each magnet ramp the damage increased further. Therefore, the complete fixation and embedding of the *RE*BCO tapes is mandatory for the TSTC. Possible options are soft solders, resins or compression inside a conduit. Each of the following samples comes with its own solution, often as Cable in Conduit Condcutor (CICC) with a copper jacket.

6.2.2 TSTC and STC sample descriptions

Within this thesis various twisted stacked tape cables (TSTC) and stacked tape cables (STC) samples were measured inside the FBI facility. Since any measured TSTC sample has its own characteristics, each sample is primarily described in the following.

16 tape STC sample A

The first sub-size cable sample, sample A, was provided by *École polytechnique fédérale de Lausanne* (EPFL). A stack of sixteen 4 mm wide *RE*BCO tapes were soldered into two copper half-shells. Main advantage of the copper casing is the increased electro-mechanical stability of the sub-size cable. Voids were completely filled with SnPb solder to avoid tape movement and movement of the copper half-shells against each other. Figure 6.21 schematically demonstrates the cable design.

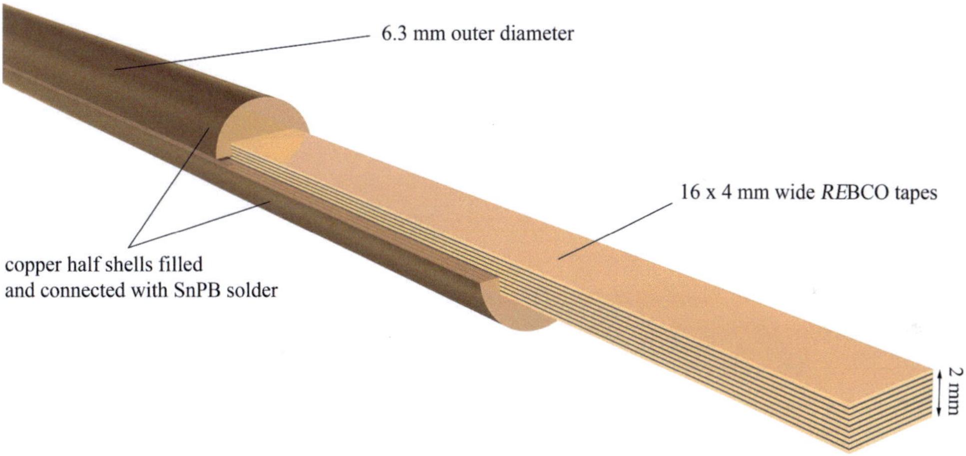

Figure 6.21: Schematic of the 16 tape stacked tape sample A inside its copper conduit which consists of two copper half-shells. The hole composition of tapes and copper half-shells is soldered together.

The measured sub-size cable was a non-twisted version of the TSTC. During the in-field measurement, the sample was oriented with its c-axis parallel to the magnetic field lines, to experience the worst case scenario with the Lorentz force pressing the superconductor tape against their edges. The sample had an overall length of 1160 mm, including two 160 mm long rectangular copper terminations. Four pairs of voltage taps were soldered along the cable in distances of 150, 250, 500 and 600 mm - each pair symmetrical to the sample centre, embracing the magnet high-field zone. A detailed sample description can be found in table 6.3. Before mounting the sample inside the FBI measurement facility, the sample was equipped with Cernox^{TM} temperature sensors and mounted into a G10 heater module, which at the same time protects the sample against Lorentz forces. Details and explanations on the heater module can be found in chapter 5, subsection 5.9.

4+4 tape STC sample B

The second sub-size cable, sample B, was manufactured at the *Karlsruhe Institute of Technology*. It shares the same tape stack compaction principle as the TSTC provided by EPFL, but comes with a slight variation in the tape cross-section. Instead of tapes of a single width, the TSTC manufactured at KIT [WFB+16] consists of two different tapes with different widths, which are arranged in form of a cross inside a electro-mechanically stabilising, solder-filled copper tube. This increases the ratio of superconductor to conduit from 0.618 to 0.787 [WFB+16]. This sample was a first, non-twisted version of the TSTC with increased conductor ratio. A schematic view of the sample is shown in figure 6.22.

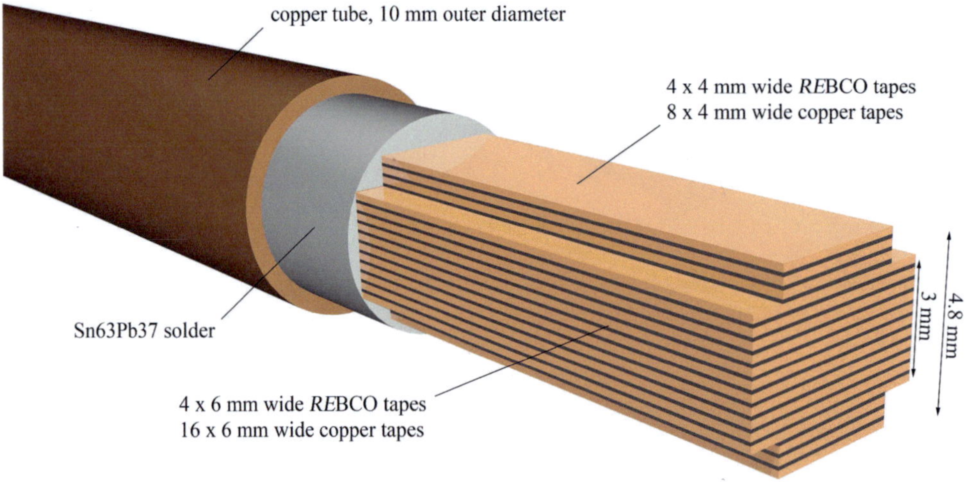

Figure 6.22: Schematic of the 4+4 tape stacked tape sample B inside its copper conduit which consists of a solder-filled copper tube. The cross design increases the ratio of superconductor to electro-mechanical stabilisation.

The tested sample consists of four 6 mm wide and four 4 mm wide *RE*BCO tapes and sixteen 6 mm wide and eight 4mm wide copper tapes which were soldered inside the copper tube with Sn63Pb37 solder. The *RE*BCO tapes were distributed equally within the soldered stack. Three pairs of voltage taps were soldered to the sample symmetrically around the high-field zone in distances of 100, 300 and 500 mm. The sample was embedded inside a G10 former to prevent it from bending due to Lorentz forces. It was mounted with the *c*-axis of the superconductor tapes in 45° to the magnetic field lines of the FBI magnet. Due to a slight twist during the manufacturing process, the orientation of the sample to the magnetic field could only be estimated after the FBI measurement by cutting the sample into half. To extend the regular $I_c(B)$-function measurement by an $I_c(\varepsilon, 12\,\mathrm{T})$-function, two CryoMak extensometers [Nyi06], that can measure longitudinal strain over a distance of 480 mm, were attached to the sample between the copper terminations. The summarized sample parameters can be found in the synoptic table 6.3.

18 tape STC sample C

The third sub-size cable, sample C, was provided by *Agenzia nazionale per le nuove tecnologie, l'energia e lo sviluppo economico sostenibile* (ENEA). It is a circular full-size cable that consists of an aluminium former and an aluminium conduit. The former contains five grooves, allocated in 120° angles around the centre axis, each of 4.3 mm width, embedding five 4 mm wide tape stacks. Each stack of superconductors is covered with four copper rods, which buffer the pressure applied by the aluminium conduit on the superconductors. Before compaction, the composition of former, superconductors and rods was wrapped with steel foil. A helium channel in the centre of the cylindrical former provides additional heat exchange between the conductors and the cooling medium liquid helium. Figure 6.23 demonstrates the cable design schematically. Details on the manufacturing process can be found in [ADC+15]. Since the current of the FBI measurement facility is limited to 10 kA, only one grove inside the former was filled with eighteen 4 mm wide *RE*BCO tapes. All other groves were filled with stainless steel tapes. Two pairs of voltage taps were used during the measurement. One with a voltage tap distance of 480 mm and one with a distance of 1300 mm, measuring the voltage drop over the superconducting niobium-titanium current leads of the FBI facility. Both pairs of voltage taps embrace the magnet high field zone symmetrically. Detailed information on the cable and superconductor parameters are found in table 6.3. The cable was embedded inside a G10 mechanical stabilisation to protect the cable against bending inside the magnetic field. Afterwards it was mounted with the *c*-axis of the *RE*BCO tape stack perpendicular to the field lines of the FBI magnet.

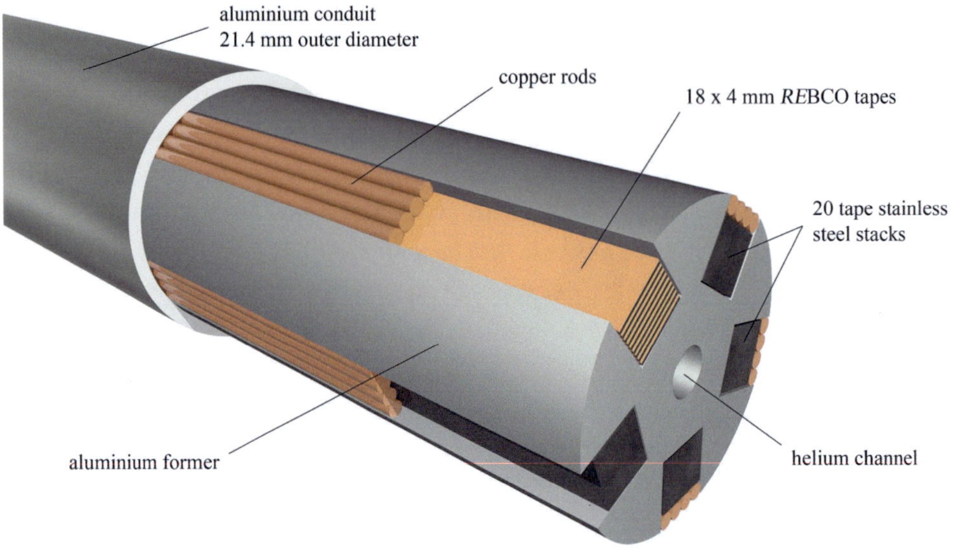

Figure 6.23: Schematic of the 18 tape full-size cable sample C. Only one of the five groves was equipped with *RE*BCO high temperature superconductors.

3×(4+2) TSTC sample D

Sample D was manufactured at the *Karlsruhe Institute of Technology*. Of all measured samples within this thesis, it was the first twisted version, which consists of three twisted super-strands that were soldered in parallel orientation inside a copper conduit. Each superstrand is manufactured of tapes of two different widths, according to the technique developed for the foregoing stacked tape cable sample B. For details on the manufacturing process I refer to [WBF+16]. A schematic visualisation of sample D is demonstrated in figure 6.24.

Sample D consists of three super-strands and three copper rods. The copper rods stabilise the super-strands and prevent the strands from moving or bending inside the copper conduit. Each super-strand consists of a copper tube with an outer diameter of 9.1 mm that contains four 6 mm and two 4 mm wide *RE*BCO tapes and sixteen 6 mm and eight 4 mm wide copper tapes. The *RE*BCO tapes are distributed equally over the entire tape stack. Each stack was soldered with Sn63Pb37 soft solder and is twisted along the longitudinal centre axis with a twist pitch of 650 mm. The super-strands are soldered with In52Sn48 solder inside the copper conduit. Four pairs of voltage taps were attached to measure the voltage drop along the cable. One pair was soldered to each super-strand in a distance of 300 mm around the magnet high-field zone. The fourth pair was attached to the cable terminations to measure the voltage drop for the entire cable over a distance of 910 mm. To protect the sample against bending inside the magnet high field zone, it was stabilised with a G10 former. Further technical details of sample D can be

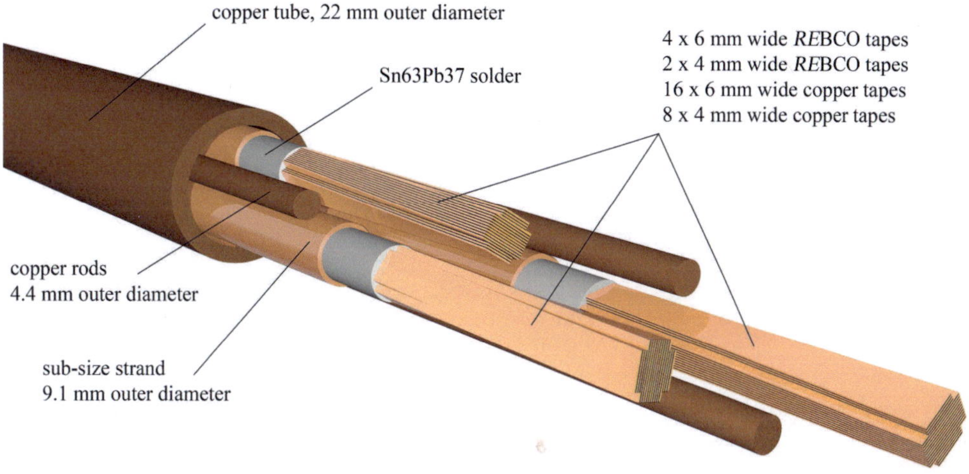

Figure 6.24: Schematic of the 3x(4+4) full-size TSTC sample D. Three TSTC super-strands, which consist of 6 mm and 4 mm wide *RE*BCO and copper tapes, are soldered in parallel orientation into a copper conduit. Additional copper rods prevent the super-strands from moving or bending inside the 12 T magnet background field.

found in table 6.3. After the sample was mounted inside the FBI facility, the tapes of the twisted strands were oriented with the c-axis in an angle of 30° to 60° to the magnetic field lines. Since the sample consisted of three strands, each twisted, the orientation could only be found out after the FBI measurement, by sawing the sample into half.

16 tape TSTC sample E

The second measured TSTC was provided by the *École polytechnique fédérale de Lausanne* (EPFL) and is the super-strand model, as it was used inside a 60 kA coated conductor cable prototype for fusion magnets, that was recently measured in the European Dipole (EDIPO) facility [UBS+15]. It is demonstrated in figure 6.25. The FBI measurements allow to compare the 60 kA EDIPO test results with the super-strand performance. The super-strand was based on the foregoing prototype sample A, with the addition of a twist around its longitudinal centre axis. Sample E consists of sixteen 4 mm wide stacked *RE*BCO tapes, that were twisted with a very tight twist pitch of only 320 mm and afterwards soldered with Sn60Pb40 soft solder into two copper half shells. The entire sample was equipped with three pairs of voltage taps at distances of 160, 320 and 1300 mm around the high-field magnet zone. The outermost pair was attached to the superconducting niobium-titanium current leads of the FBI facility. Since this sample was investigated for its temperature dependant $I_c(B)$-function, the entire strand was embedded inside the G10 heater module, described in chapter 5, subsection 5.9. As most samples, Sample E was mounted inside the FBI facility with the c-axis of the superconductor tapes parallel to the magnetic field. Technical details on the sample and the superconductors are listed in table 6.3.

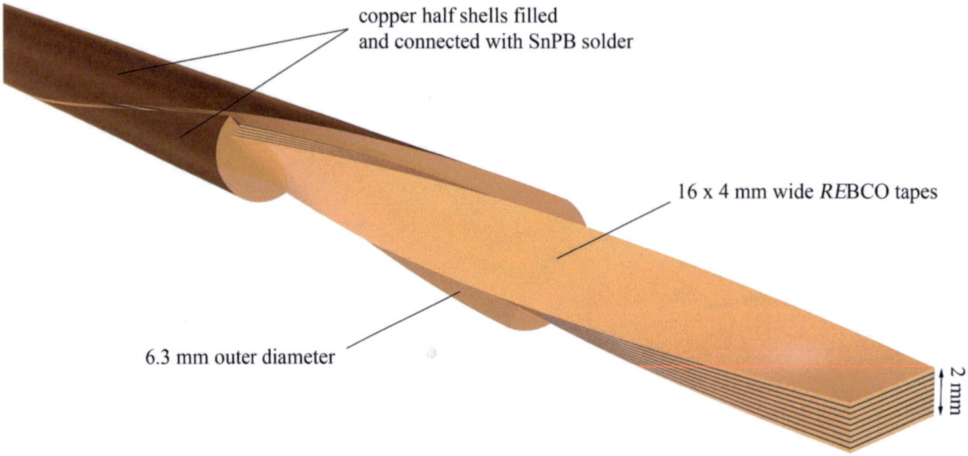

Figure 6.25: Schematic of the 16 tape TSTC sample E. The 4 mm wide superconductors are soldered into two copper half-shells and were afterwards precisely twisted. This sample is the super-strand model used for a 60 kA coated conductor cable prototype for fusion magnets, that was measured in the EDIPO facility.

32 tape TSTC sample F

Last investigated TSTC sample is sample F, which was manufactured at the *Karlsruhe Institute of Technology*. As the foregoing models, it was manufactured of *RE*BCO tapes of two different widths. The main difference to its predecessors is the full equipment with superconductors. Copper spacer tapes were not implemented. The whole cable was twisted along its longitudinal centre axis with a minimum twist pitch of 300 mm. A model of the sample F is shown in figure 6.26. The tape stack was manufactured with 22 6 mm wide and 10 4 mm wide *RE*BCO tapes. It was soldered to a stack, wrapped with Sn63Pb37 solder wire and afterwards compacted inside an oxygen-free high thermal conductivity (OFHC) copper tube. The soft solder was not heated afterwards any more. As the thermal conductivity of the OFHC copper is very high, the cable centre, which is in the high-field zone of the magnet, was thermally isolated by cutting 0.5 mm inside the copper. Furthermore, copper and solder was removed to fix the temperature sensors directly on the superconducting tapes. The final composition was embedded into an adjusted G10 heater module, described in chapter 5, subsection 5.9. Three pairs of voltage taps were soldered to the sample in distances of 160, 628, 895 mm. Technical details of sample F are summarized in table 6.3.

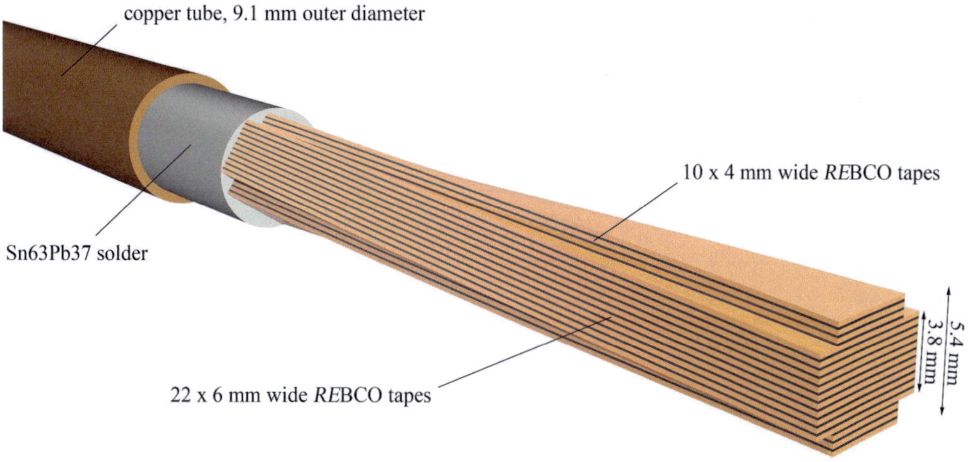

Figure 6.26: Schematic of the 32 tape TSTC sample F. The 4 mm and 6 mm wide superconductors are stacked, wound with solder wire, compacted inside a copper tube and afterwards twisted around its longitudinal axis.

Table 6.3: Parameters of the measured STC and TSTC samples.

parameter	sample A: 16 tape STC	sample B: 8 tape STC	sample C: 18 tape STC
provided by	*École polytechnique fédérale de Lausanne*	*Karlsruhe Institute of Technology*	*Agenzia nazionale per le nuove tecnologie, l'energia e lo sviluppo economico sostenibile*
sample length (mm)	1160 (incl. terminations)	1160 (incl. terminations)	1160 (incl. terminations)
superconductor	SuperPower SCS4050-AP	SuperPower SCS6050-AP, SCS4050-AP	SuNAM SCN04150
production batch	M4-157-2 FS	M3-936-1 FS (6 mm), M4-69-4 BS (4 mm)	140127-03
superconductor tapes	16×4 mm	4×6 mm, 4×4 mm,	18×4 mm
tape $I_c(77\,K)$	102 A	166 A, 114 A	225 A
additional spacer	-	16×6 mm, 8×4 mm Cu tapes	120×4 mm SS tapes, 20 Cu rods
stack height (mm)	2	4.8	2.9
copper stabilisation (μm)	50	100	40
substrate (μm)	50	50	100
twist pitch (mm)	untwisted	untwisted	untwisted
conduit	two solder filled Cu half-shells	solder filled Cu tube	compacted aluminium conduit
solder	Sn60Pb40	Sn63Pb37	Sn63Pb37
termination length (mm)	160 each	300 each	230 each
outer cable diameter (mm)	6.3	10	21.4
pairs of voltage taps	4	3	2
voltage tap distances (mm)	150, 250, 500, 650	100, 300, 500	480, 1300
mechanical stabilisation	G10 heater module	G10 former	G10 former
extensometer distance (mm)	-	480	-
in-field tape orientation	$\measuredangle\left(\vec{B}, c\text{−axis}\right) = 0°$	$\measuredangle\left(\vec{B}, c\text{−axis}\right) = 0°$	$\measuredangle\left(\vec{B}, c\text{−axis}\right) = 0°$

Table 6.4: Parameters of the measured TSTC samples.

parameter	sample D: 3×(4+2) tape TSTC	sample E: 16 tape TSTC	sample F: 32 tape TSTC
provided by	*Karlsruhe Institute of Technology*	*École polytechnique fédérale de Lausanne*	*Karlsruhe Institute of Technology*
sample length (mm)	1160 (incl. terminations)	1160 (incl. terminations)	1150 (incl. terminations)
superconductor	SuperPower SCS6050-AP, SCS4050-AP	SuperOx 4 mm tape	SuperPower SCS6050-AP, SCS4050-AP
production batch	M3-936-1 FS (6 mm), M3-1048-2 0508 (4 mm)	2014-14-L	M4-249-2 0712 (6 mm), M3-1062-2 0710 (4 mm)
superconductor tapes	3×4×6 mm, 3×2×4 mm	16×4 mm	22×6 mm, 10×4 mm
tape $I_c(77\,K)$ (A)	171, 119	155	184, 144
additional spacer	3×16×6 mm, 3×8×4 mm Cu tapes	-	-
stack height (mm)	3×5.2	1.9	5.4
copper stabilisation (μm)	100	40	100
substrate (μm)	50	60	50
twist pitch (mm)	650	320	300
conduit	3 solder filled Cu tubes, Cu conduit	two solder filled Cu half-shells	wrapped solder inside copper tube
solder	Sn63Pb37 strands, In52Sn48 cable	Sn60Pb40	Sn63Pb37
termination length (mm)	125 each	160 each	120 each
outer cable diameter (mm)	22	6.2	9.1
pairs of voltage taps	4	3	3
voltage tap distances (mm)	300, 300, 300, 910	160, 320, 1300	160, 628, 895
mechanical stabilisation	G10 former	G10 heater module	G10 heater module
extensometer (mm)	-	-	-
in-field tape orientation	$\measuredangle\left(\vec{B}, c\text{-axis}\right) = \sim 45°$	$\measuredangle\left(\vec{B}, c\text{-axis}\right) = 0°$	$\measuredangle\left(\vec{B}, c\text{-axis}\right) = 0°$

6.2.3 Field dependant measurements

Research and development of the investigated HTS samples is an iterative process. Therefore, the samples were measured one after another over the entire period of this PhD thesis. Similarities in the preparation and the measurement process, that account for all samples, will only be discussed once and not for each sample individually.

Durability against Lorentz forces

For each sample, the first investigation was the measurement of possible degradation due to the high Lorentz forces inside the magnet background field at 4.2 K. Since the $I-B$-curve of high temperature superconductors does not drop linearly, but exponentially, the maximum Lorentz force is expected at the peak magnetic field of 12 T. Therefore, it is not reasonable to acquire the first $I_c(B)$-function at 12 T and take the risk of damaging the cable during the first measurement. However, it is neither reasonable to begin the first measurement at very low magnetic fields or even self-field, since the maximum critical current of the sample increases inverse proportional to the magnetic field. If the sample quenches while carrying high currents, the electro-mechanical stabilisation of the superconductors and the cable might be insufficient to carry the current and prevent the cable from fusing. An irreversible cable damage would be the result. In consequence, all sub-size cable investigations were started at magnetic fields between 6 T and 10 T. The chronological order of the measurement process, including the magnet ramps, is visualized in each plot with arrows. The results of the 4.2 K in-field measurements are plotted in figure 6.27 and figure 6.28.

For all samples, the magnetic field was cycled between 12 T and lower magnetic fields. Except for sample C, none of the investigated samples show degradation or training effects due to Lorentz forces inside the magnetic field. The acquired sample-individual $I_c(B)$-function remains the same during the cyclic magnetic field ramps. In consequence, these samples proved to resist their individual Lorentz forces, induced by the 12 T background field and the maximal sample specific critical current $I_c(12\,\mathrm{T})$. Sample C experienced a training effect during the first three 10 T measurements. The critical current dropped from 2.42 kA to 2.22 kA. After this initial training, no further training or degradation was observed and sample C withstood the further experienced Lorentz forces.

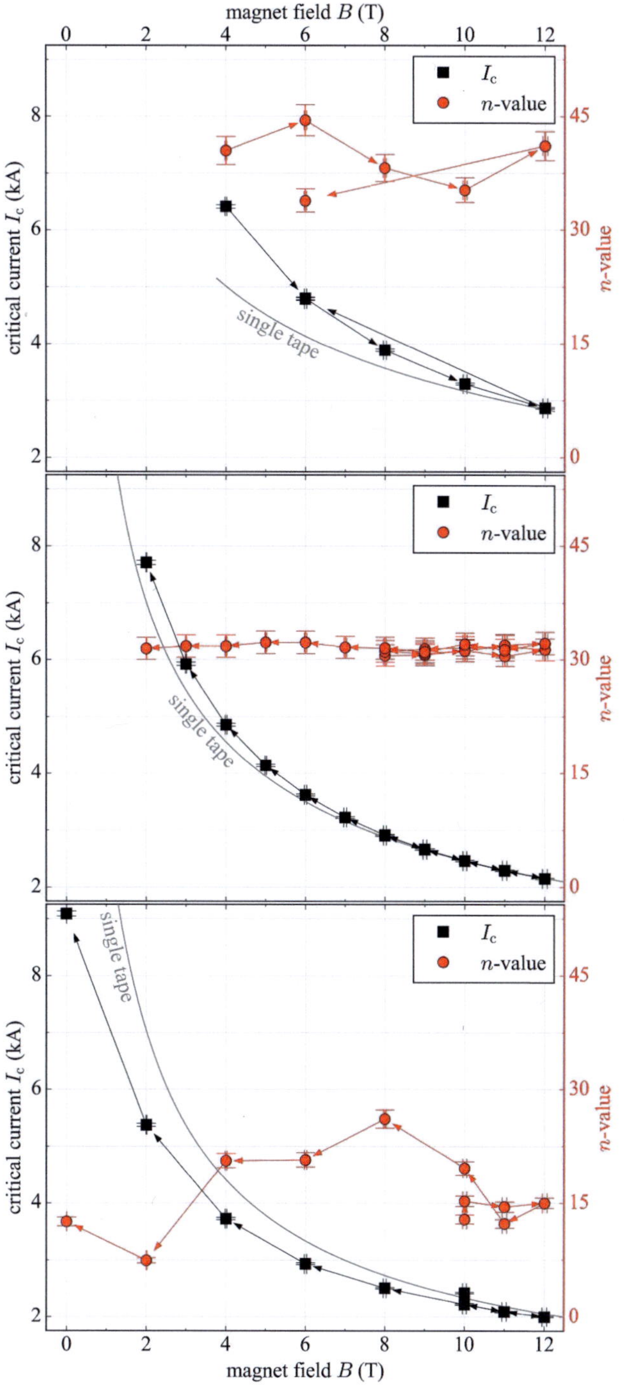

Figure 6.27: In-field performance of sample A, B and C (top down) at 4.2 K.

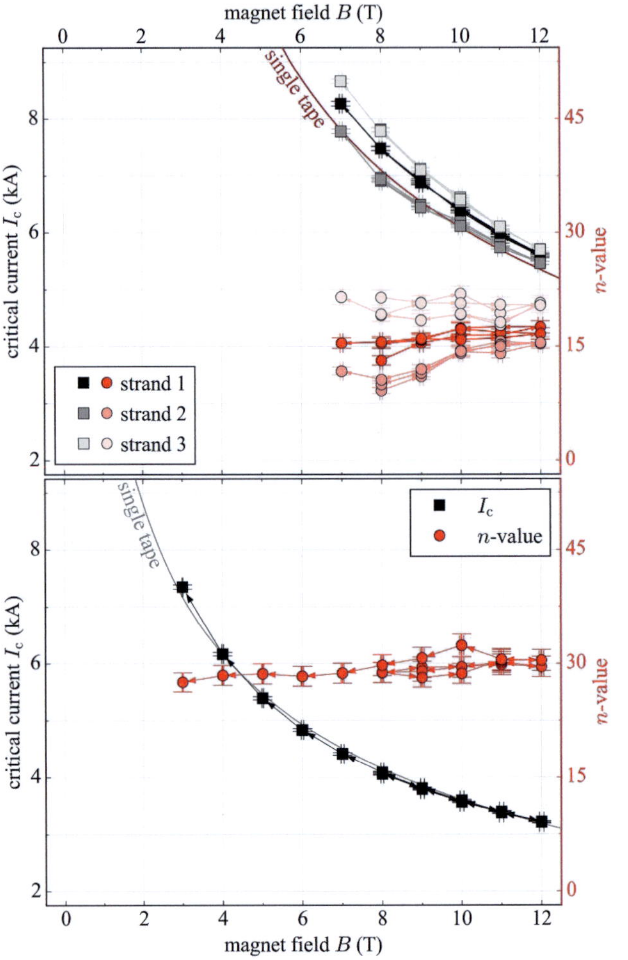

Figure 6.28: In-field performance of sample D and E (top down) at 4.2 K.

Individual performance and comparison with single tape data

Beside critical currents and n-values of the measured samples A to F, 6.27 and figure 6.28 also include the single tape data of the superconductor tapes they were manufactured with. The performance of each sample will be discussed individually.

For sample B, C and D single tape curves are calculated from the number of superconductor tapes per sample m, the critical tape current in liquid nitrogen $I_c(77\,K)$ and the field-dependant lift-factor $f_l(B)$ by

$$I_{\text{tape}}(B) = m \cdot I_c(77\,K) \cdot f_l(B).$$

For sample A and E, the single tape curves are calculated by equation

$$j_c(B,T) = A \cdot \frac{B_{irr}(T)^{\beta}}{B} \cdot \left(\frac{B}{B_{irr}(T)}\right)^{p} \cdot \left(1 - \frac{B}{B_{irr}(T)}\right)^{q}, \; B_{irr}(T) = B_{irr0} \cdot \left(1 - \frac{T}{T_c}\right)^{\alpha} \tag{6.2}$$

with the fit parameters α, β, p, q, the irreversibility field $B_{irr}$and the critical temperature T_c [UBS+15].

- Sample A: the measured $I_c(B)$-function of STC sample A lies about 14% above the single tape data. The current density for the single tape curve was calculated by equation 6.2. Since the performance of superconductor tapes varies with the production batch, and the *RE*BCO tapes of sample A are not from the same batch as the tapes used for the single tape current estimation, both curves do not overlay. However, for magnetic fields above 10 T, the current of sample A coincides with the single tape. With decreasing magnetic field, the spread between both curves increases. Sample A exceeds the single tape data. The n-values of sample A are, with the exception of the 6 T measurement, in accordance with typically expected values between 30 and 40. Both, the current curve and the stable, high n-value indicate a well-made HTS cable sample.

- Sample B: the measured $I_c(B)$-function of STC sample B is almost perfectly congruent with the single tape data. Again, the lift-factor data was taken from [Sup15]. The sample behaved as it would have been expected by a single tape. The constant n-values above 30 indicate a steep superconducting transition for all measured magnetic fields. As sample A, it is a very good HTS sub-size cable with promising in-field characteristics.

- Sample C: the measured $I_c(B)$-function of sample C lies for high magnetic fields about 5-10% below the single tape data. The lift-factor for the single tape was taken from [DCA+15]. The reason for this margin is an 8% degradation due to Lorentz forces, which occurred during the first two 10 T measurements. The *RE*BCO tapes were pressed with their edges against the aluminium former. As already seen in subsection 6.1.2, such pressure direction damages the tapes permanently. This cable was the first HTS full-size cable design, that was measured in-field up to that time. The small degradation plays a minor role and can be reduced by increasing the transversal compaction pressure as suggested for the RACC in subsection 6.1.3. This sample, among others, proves the STC respectively TSTC to be a valid candidate for large scale magnet cables.

- Sample D: the measured $I_c(B)$-function of sample D is difficult to compare with the single tape data. After the FBI measurement the sample was cut in half, to estimate the angle between the superconductor c-axis and the magnetic field lines. It was found, that the angle varied between 30° to 45° because each twisted strand had a slightly different orientation. The lift-factor for single tapes was taken from [Sup15]. Since the single tape in-field performance is for 45° almost the same as for 0°, the single tape curve for 0° can be used as indication. Although all three strands have the same number of superconductor tapes, their critical current varies. Strand 1 and strand 2 must have slightly be damaged during the twisting or compaction process. Training- or degradation effects due to experienced Lorentz forces were not observed. The n-values between 10 and 22 are comparably low for a twisted stacked tape solution. However, for this sample type, both twist and compaction were first time tested features. Considering these facts, the sample is a promising first approach which leaves space for further optimisation.

- Sample E: another sample with an $I_c(B)$-function, which is perfectly congruent to the single tape data. The critical current density for the singe tape was calculated with equation 6.2. The n-value remains constant around 30 with a very slight decrease towards lower magnetic fields. Sample E is the first twisted strand, which shows exactly the same characteristics as expected by a single tape. Neither the manufacturing nor the twisting procedure impaired the current carrying capabilities of the *RE*BCO superconductor tapes. This sample is a fully suitable sub-size strand.

- Sample F: Due to its high critical current, sample F could not be measured at 4.2 K. Even at the maximum current of 10 kA a superconducting transition could not be observed. The temperature dependent measurement follows in next section.

6.2.4 Temperature dependant measurements

During a quench inside a fusion toroidal field coil cable, the superconductor temperature might exceed 120 K respectively 220 K, depending on the quench detection voltage of 100 mV or 400 mV [HGF+16]. To calculate the temperature evolution during and after the quench it is important to know the current carrying capabilities of possible cable candidates. Therefore, several samples were investigated for their temperature and field dependant critical current $I_c(B,T)$. Heater modules, individually adjusted to the sample dimensions, were used to increase the sample temperature inside the liquid helium bath. Details on the heater module are found in subsection 5.9. The temperature and field dependant measurements were executed for sample A, sample E and sample F. All results are visualized and compared with the single tape data in figure 6.29. Since it is not possible to precisely propagate the uncertainty of the temperature measurement to the critical current without having a precise $I_c(T)$-function, the uncertainty $u_c(T)$ is added to the single tape data and marked with a red area. For all following temperature dependant measurements, the field and temperature dependant single tape curves were calculated with equation 6.2.

- Sample A: For all measured temperatures, the critical sample current slightly exceeds the critical single tape current. This effect increases with decreasing magnetic field. In principle, sample A shows the expected $I_c(B,T)$-function. The n-value of sample A is not constant for any single temperature during magnet ramp. With increasing temperature, the n-value drops from values above 30 to values between 8 and 30. With decreasing magnetic field, the n-value of the sample increases. Large deviations from the expected single tape values were not observed.
- Sample E: in comparison to sample A, the $I_c(B,T)$-function of sample E is congruent with the calculated single tape data for the 4.2 K, 8.5 K and 17 K measurements. For lower temperatures, the sample current slightly exceeds the expected single tape data. Measured n-values drop with increase of the sample temperature. The lowest measured value was six at a sample temperature of 44 K. For each sample temperature, the n-value is independent of the magnetic field and remains on a constant level. Sample E satisfied all expectations of the temperature dependant measurement.
- Sample F: the measured temperature dependency was far from expected single tape data. For low temperatures, the critical current values of sample F exceed the single tape data by far. With increase of the temperature, the cable data approaches the single tape data and only for temperatures above 50 K the cable data matches the theoretical expectations. This effect comes from the OFHC copper. For temperatures below 50 K, its thermal conductivity rises and peaks at 20-25 K with a factor of 5 compared to its thermal conductivity at 50 K. Regular Electrolytic-Tough-Pitch (ETP) copper peaks with a factor of 2-3. The reduction of the thermal conductivity by cutting the copper tube was not sufficient.

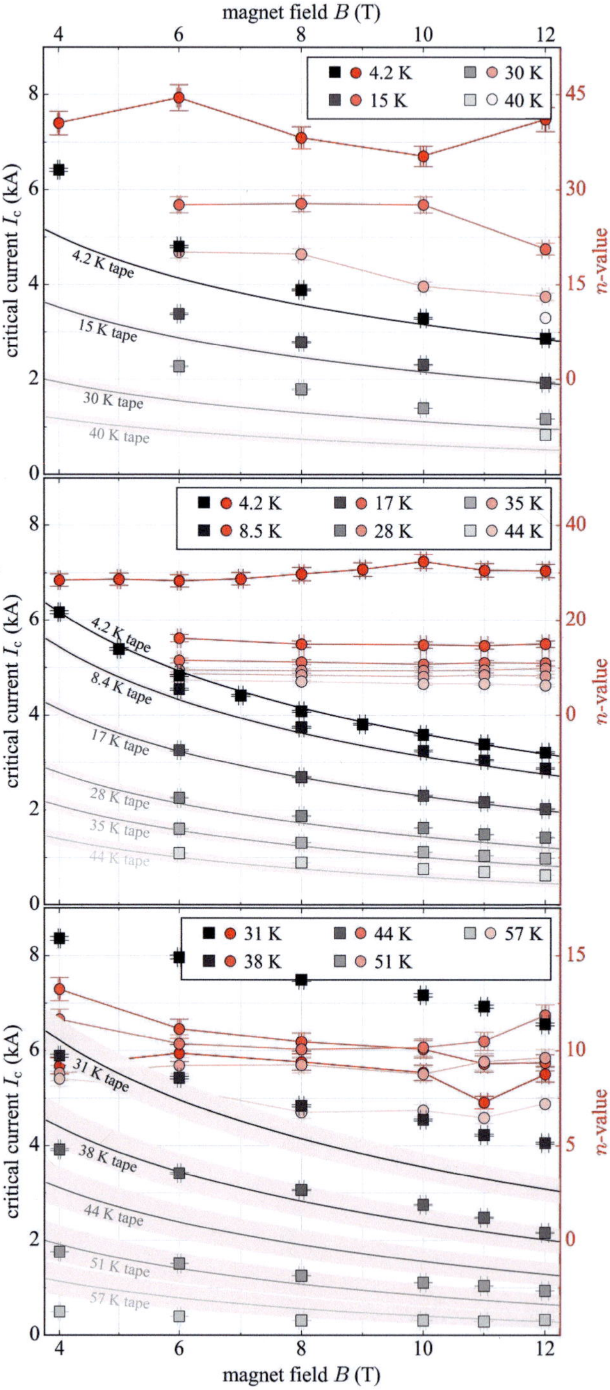

Figure 6.29: In-field performance of sample A, E and F (top down) at various temperatures.

6.2.5 Strain dependant measurement

Lorentz forces, acting on the superconductors of the magnet toroidal field coil, result in hoop stress. Due to the large diameter of the DEMO toroidal field coil and its comparably thin winding-pack thickness, the superconductors experience hoop stress mostly as stress in longitudinal direction. With increasing cable elongation, the critical current of the cable drops significantly. Therefore, it is important to understand the impact of longitudinal strain on a sub-size cable strand. For this reason, sample B was prepared for a tensile test and mounted to the FBI facility. The tensile strain was measured with two extensometers. Technical descriptions can be found in subsection 6.2.2. Since any tested sample is irreversibly damaged during the strain test, the investigation was executed at a background field of 12 T, to acquire the $I_c(\varepsilon, 12\,\mathrm{T})$-function. The results of the measurement are plotted in figure 6.30.

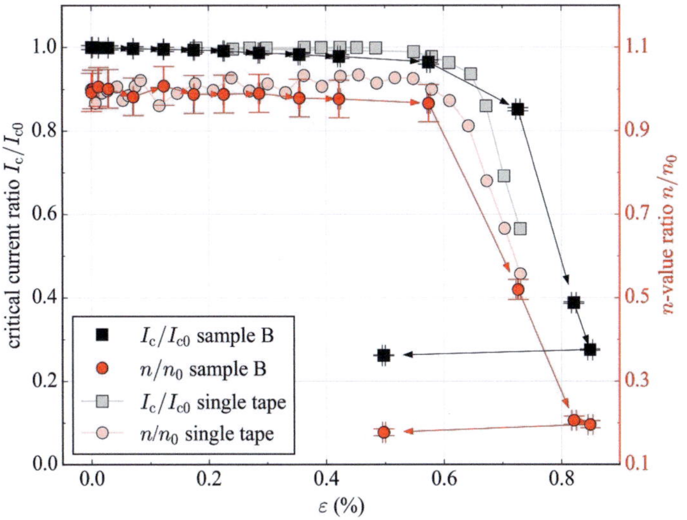

Figure 6.30: Critical current ratio and n-value ratio of sample B, depending on the longitudinal cable strain. The measurement was performed at 4.2 K inside a 12 T background field. Arrows show the measurement chronology. The single tape data is taken from [Bar15].

For strain below 0.4%, both critical current and n-value of sample B remain constant. For strain between 0.4% and 0.6%, both values drop by approximately 4%. After exceeding a strain of 0.6%, the cable degrades rapidly. At strain values of 0.8%, the critical current already dropped by 60% and the n-value by 80%. The last data point is taken at a strain of 0.5%. Since critical current and n-value do not recover, sample B degraded irreversibly. Comparing the FBI measurement results with single tape data provided by [Bar15], it can be seen, that sample B almost perfectly behaves as the single tape. The critical currents and especially the n-values concord.

During the tensile test, the tensile load was measured with a 100 kN load cell, implemented in the tensile machine. The load-strain and the stress-strain curve for the sample B measurement are demonstrated in figure 6.31.

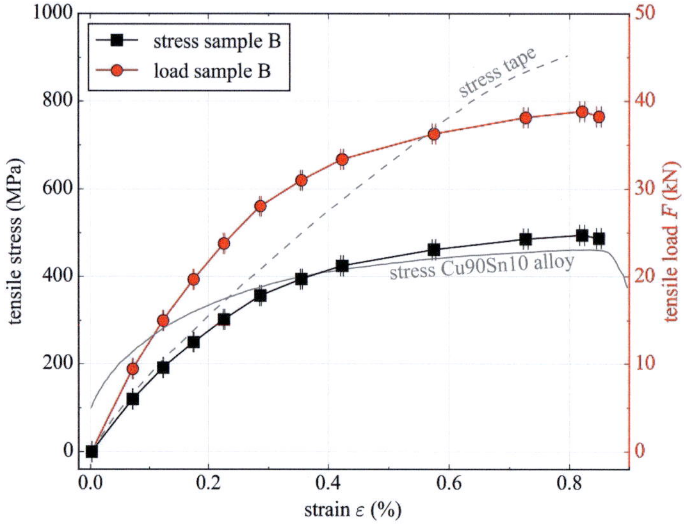

Figure 6.31: Load-strain and stress-strain diagram for sample B. The measurement was performed at 4.2 K inside a 12 T background field. The single tape data is taken from [Bar15, Cop].

The tensile stress of sample B was calculated dividing the measured load by the cable cross-section area (πr^2). The single *RE*BCO tape data was provided by [Bar15]. For stresses below 200 MPa, sample B and the single tape share the same strain values. For larger stress values, the single tape strain exceeds the sample B strain by far. This margin comes from the material composition of the sub-size cable. While in a single *RE*BCO tape the stainless steel substrate dominates above any other materials, in the sub-size cable, however, the electro-mechanical copper conduit and the Sn63Pb37 solder occupy more than 70% of the cross-section area. Therefore it is not surprising, that the stress-strain curve of sample B is very close to the stress-strain curve of commercially available, annealed bronze (Cu90Sn10) alloy. The data for the bronze alloy is taken from [Cop]. In consequence, the current-strain behaviour for STC sub-size cables can be derived from *RE*BCO single tape data. For a stress-strain estimation however, the focus should lie on the stabilising materials such as copper and solder alloys, which will always be dominant for HTS full-size toroidal field coil cables, where a high copper ratio is necessary to provide sufficient electrical stabilisation.

6.2.6 Conclusion and discussion

During several FBI in-field measurements, the Stacked Tape Cable (STC) and the Twisted Stacked Tape Cable (TSTC) have proven to be promising concepts for sub-size strands in full-size toroidal field coil cables. They have a high durability against Lorentz forces, almost perfectly full-fill the single tape expectations in temperature dependant measurements and benefit from the high *RE*BCO tape durability against longitudinal strain. Further discussion and the comparison with other cable concepts continues in sub-section 6.4.

6.3 Conductor on Round Core (CORC)

The Conductor on Round Core (CORC) is a cable with semi-transposed 4 mm wide *RE*BCO tapes, which are wound in an angle of 45° around a central former, usually a copper tube. While the first layer only consists of two tapes, the number of tapes increases with the cable diameter. Each layer is wound in the opposite direction as the layer below. Since the tapes are not soldered but can move against each other, the finished cable is laminated into a thin foil to prevent tape movement, while keeping it flexible. The CORC cable was developed at the University of Twente, Netherlands, and is now produced by Advanced Conductor Technologies which provided the investigated sample.

6.3.1 Mechanical properties

Main advantage of the CORC cable is its flexibility, as it can be bent to radii of few centimetres [LLJ+13, WLM+16]. With tapes wound around a central former, the CORC does not have a preferred tape orientation and therefore no preferred bending direction. Furthermore, the CORC cable is expected to prove a high resilience against tensile strain. Due to the 45° winding, longitudinal tensile strain is applied 45° to the *a* and *b*-axis of the *RE*BCO lattice, which is the direction of highest resilience against tensile strain [LAP+11]. The expectations are to be verified in concluding strain measurements. Compared to a conventional TSTC, the 45° winding of the CORC increases the required tape length per cable length by a factor of $\sqrt{2}$. This factor can be reduced by reducing the winding angle. However, lower winding angles imply lower bending radii and lower resilience against tensile strain.

6.3.2 Sample description

The investigated CORC cable consists of 24×3 mm and 21×4 mm wide *RE*BCO tapes which are wound around a 5 mm copper former in 12 layers resulting in an outer diameter of 7.6 mm.

After the winding process, the outer cable diameter is 7 mm. The maximum twist pitch is 33.8 mm. Figure 6.32 demonstrates the cable design. As the sample is cylindrically wound around the former, there is no preferred sample orientation during the in-field measurement. The sample has an overall length of 1160 mm, including two 300 mm long rectangular copper terminations. Each of the three pairs of voltage taps is attached to one bundle of *RE*BCO tapes. The distance for each voltage tap is 56 cm. A detailed sample description can be found in table 6.5.

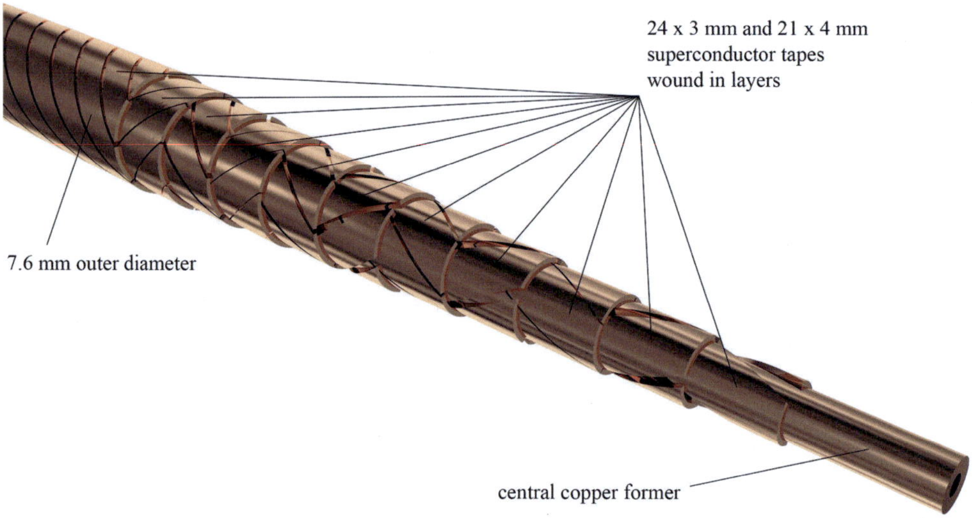

Figure 6.32: Schematic of the 24 tape CORC cable. The 4 mm wide superconductors are wound around a copper former and afterwards laminated with a thin foil.

6.3.3 Field dependent measurement

The field dependent measurement could not be terminated successfully. During the first 10 T measurement the cable quenched at a current of 9.4 kA. The critical current criteria was not even reached. Post investigation suggests that the G10 cable reinforcement did not withstand the Lorentz forces and allowed for cable movement of few millimetres, which lead to a spontaneous cable quench. Although the quench detection immediately triggered the interlock of the power supply, the cable fused. Inside the magnet high-field zone, where the cable is completely surrounded by the copper plates of the G10 heater module (section 5.2.5), the cable was not damaged at all. However, few millimetres beyond both ends of the copper contact plates the cable was completely vaporized. It is assumed, that the copper of the contact plates provided sufficient thermal and electrical stability to protect the cable inside the 15 cm long plates. The

Table 6.5: Parameters of the measured CORC sample.

parameter	24+21 tape CORC
provided by	*Advanced Conductor Technologies*
sample length (mm)	1160
superconductor	SuperPower
superconductor tapes	24×3 mm, 21×4 mm
tape I_c(77 K)	80-90, 100-130
copper stabilisation (μm)	40
substrate (μm)	50
maximal twist pitch (mm)	33.8
termination	solder filled Cu terminations
termination length (mm)	300 each
outer cable diameter (mm)	7.6
pairs of voltage taps	3
voltage tap distances (mm)	560, 560, 560
mechanical stabilisation	G10 heater module

inherent electromechanical stability of the CORC, however, is very low and was insufficient to prevent a complete burnout outside the copper stabilisation.

Furthermore, the QD voltage was taken from one of the three 56 cm voltage taps, that were directly soldered to three *RE*BCO tape bundles. In case these bundles quench time-displaced, the voltage tap of the QD might have been last to measure the quench. Due to the fact, that the CORC has a very bad current sharing between the *RE*BCO layers, valuable time would pass until a quench is measured with all three voltage taps. For future measurements the QD will be attached to the cable terminations instead of a single superconductor bundle. Still, it is doubtful that irreversible cable damage could have been prevented during the cable quench.

6.3.4 Conclusion and discussion

The CORC sample was destroyed during measurement at 10 T. Bad current sharing between the *RE*BCO tapes and the very low copper ratio makes it rather difficult to protect the cable from burnout during quench. At currents of almost 10 kA the dissipated heat is enormous and requires sufficient thermal stability, which is not provided by helium bath cooling. The low copper ratio inside a CORC further increases local heat generation. In comparison to a copper stabilised STC or TSTC, which is difficult to destroy by over-current, a non-stabilised CORC is vulnerable to dissipated heat and over-current.

6.4 Summary and evaluation

Within this thesis, nine different high temperature superconductor sub-size cables have been prepared, mounted and finally measured inside the FBI facility at different magnetic fields and temperatures: two Roebel Assembled Coated Conductors (RACC), three Stacked Tape Cables (STC), three Twisted Stacked Tape Cables (TSTC) and one Conductor on Round Core (CORC). Each cable comes with its own design, with different numbers of superconductors from different superconductor manufacturers and its own characteristics. The in-field curves at 4.2 K of the seven successfully measured samples is demonstrated in figure 6.33. For comparison reasons, the measured data of each sample was fitted with the perfectly matching, empiric, exponential function

$$f(x) = a \cdot e^{-bx} + c, \tag{6.3}$$

and is plotted in the range of the original measured data. Except for RACC sample A and STC sample C, which both experienced a slight degradation of 4.2% respectively 8%, all other samples proved to resist the experienced Lorentz forces. However, since each sample came with a different amount of superconductors, the experienced Lorentz forces vary from sample to sample. Therefore, the experienced in-field characteristics are summarized in table 6.6. This table contains critical currents I_c and n-values at 12 T, the peak Lorentz forces F_L, the pressure experienced by the sample p_{sample} respectively the superconductor stack p_{stack} and observed strand degradation. The Lorentz force was calculated by integration of the magnetic field in radial magnet bore direction along the 90 mm high-field zone by the empiric equation

$$F_L = 2 \cdot I_c \cdot \int \left(\frac{1}{0.08338 + 29.04224 \cdot x^{3.06406}} \right) dx,$$

which was created after magnetic field measurements. The pressure p_{sample} was calculated, dividing the Lorentz force by the contact area between the particular sample and the G10 stabiliser within the high-field zone. The pressure p_{tape} is the pressure actually experienced by the *RE*BCO tapes, calculated, dividing the Lorentz force by the area of the *RE*BCO tapes perpendicular to the Lorentz force.

The largest current at 12 T was measured for TSTC sample F with 10 kA. Since this cable was a full-size cable, the high value was expected. It is also the sample that experienced the highest absolute Lorentz force of 10.73 kN. However, the tape pressure is not higher as for STC sample A and B and TSTC sample D and E, which all show values above 20 MPa. The best n-values, with n-values above 30, were achieved for RACC sample A, STC sample A, STC sample B and TSTC sample E. After all, the samples with the smallest diameter and the highest superconductor ratio withstood the highest tape pressure values, because they had one thing in common: the *RE*BCO tapes were completely soldered inside a copper conduit.

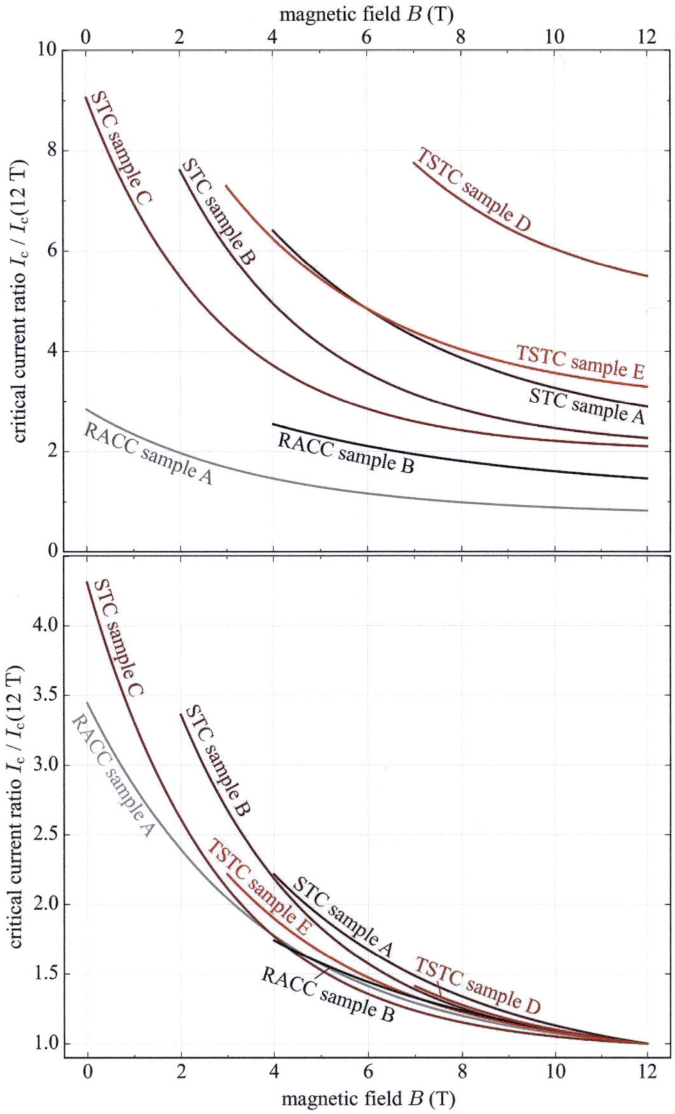

Figure 6.33: Comparison of the measured FBI samples at 4.2 K for different magnet background field values. All curves were obtained by fitting the FBI measurements with equation 6.3.

Table 6.6: Comparison of the investigated samples for critical current and n-value at 12 T, the experienced peak Lorentz force and peak pressure at the critical current criteria and the observed degradation due to the pressure. All Lorentz forces F_L were calculated by integrating the 12 T magnetic field over the 90 mm magnet high-field zone. The sample pressure p_{sample} was calculated, dividing the Lorentz force by the contact area between the entire sample and the corresponding G10 former inside the high-field zone. The tape pressure p_{tape} is the Lorentz force divided by the area of the REBCO tapes inside the particular sample.

sample	I_c(12 T) (kA)	n (12 T)	F_L (kN)	p_{sample} (MPa)	p_{tape} (MPa)	degradation (%)
RACC sample A	0.80	8	0.86	0.58	13.65	4.2
RACC sample B	1.45	34	1.56	1.05	17.33	-
STC sample A	2.87	41	3.08	3.46	21.39	-
STC sample B	2.16	32	2.32	1.64	21.48	-
STC sample C	2.00	15	2.15	0.71	9.48	8
TSTC sample D	5.70	16 - 21	6.11	1.96	24.77	-
TSTC sample E	3.22	30	3.45	3.94	23.96	-
TSTC sample F	-	-	10.73	8.34	22.08	-

For both RACC cables, the tape pressure is not equally distributed as one might expect. This results from the calculation of the stack height t_{stack}, which is

$$t_{stack} = \frac{m-2}{2} \cdot t_{tape} + 2 \cdot t_{tape},$$

with m the number of Roebel strands and t_{tape} the strand thickness. While $m-2$ strands are distributed on two parallel stacks, two strands cross-over from one stack to the other on the top and bottom side of the RACC. With increase of the number of tapes, the tape pressure of the RACC approaches the tape pressure of a stacked tape cable:

$$\lim_{m\to\infty} p_{tapeRACC} = p_{tapeSTC}.$$

To conclude the sample comparison, it is necessary to compare the current densities. Due to the tape structure of the REBCO superconductors it is convenient to calculate the critical current I_c per millimetre tape width, dividing the critical sample current I_c by the number of tapes m and the tape width w:

$$I_c/\mathrm{mm} = \frac{I_c}{m \cdot w}.$$

The resulting $I_c(B)/mm$-curves are demonstrated in figure 6.34. The curves consider the exact tape widths of each sample and the degradation effects measured during the FBI experiments.

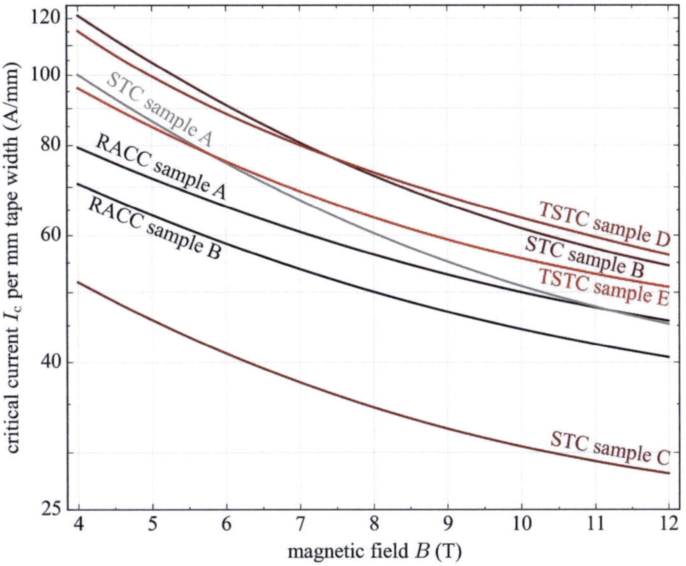

Figure 6.34: Comparison of the critical current per millimetre tape width for all measured sub-size cables at 4.2 K. All curves were obtained by fitting the FBI measurements using equation 6.3 and dividing the result by the number of tapes and the tape width.

Sample D and sample B were composed of *RE*BCO tapes with a very good in-field performance. The critical current per millimetre tape width at 12 T is around 50 A/mm sample width. STC sample C contained superconductor tapes with a comparably bad in-field performance. Even after subtracting the 8% degradation, the critical current per millimetre tape width is still only 30 A/mm.

As already discussed, all sub-size cables, except for RACC sample A and STC sample C, did not degrade inside the magnetic field and were in accordance to the single tape data. Therefore, data in figure 6.34 represents the single tape curves of the utilized *RE*BCO samples. The enormous spread between the tape curves implicates that the main aspect, which influences the current-carrying capabilities of any sub-size cable, is not the cable design, but the single tape performance. Therefore, the focus of each cable design should lie on the electro-mechanically stabilisation of the single tapes to compensate the experienced pressure per tape, which scales with the *RE*BCO tape quality. So far, all sub-size cable types, the RACC, the STC, the TSTC and the CORC proved to satisfy this claim. However, if current densities increase further for future *RE*BCO tapes, methods like compaction by applying pressure might come to their limits. Soldering the *RE*BCO tapes into a stabilising structure might become the only option to keep the required electrical and mechanical stability. For DEMO toroidal field coil cables, however, this is not of concern. Up-to-date HTS superconductors already fulfil the required current densities. Higher current densities need to be avoided, to achieve the targeted copper to superconductor ratio inside the DEMO full-size cable.

In the following chapter, the measured sub-size cable designs are confronted with the DEMO TF coil requirements. The focus lies on the applicability of the cable designs and their up-scaling to a full-size DEMO TF coil cable.

7 Conceptual design of an HTS TF coil fusion cable

All foregoing investigated samples were sub-size cables. In order to use any of these cable designs inside a toroidal field coil, it is necessary to scale the cable designs from sub-size to full-size fusion concepts, under consideration of the requirements given by the DEMO coil design, discussed in chapter 3.3. The following chapter highlights the advantages of each cable design individually and demonstrates potential cable concepts to be used in DEMO TF coils.

7.1 DEMO TF coil requirements

The requirements for the DEMO TF coil were estimated in chapter 3.3. The magnetic field of 6.82 T at the centre line of the plasma torus demands a coil current of 19.06 MA, which corresponds to a conductor current of 50 kA for 384 coil windings. Considering electromechanical stabilisation and helium forced flow, the conductor requirement is specified by its cross-section area, which is divided into superconductor, copper, stainless steel and helium void. Table 7.1 summarizes the exact division for a full-size conductor. These requirements are taken as basis for following the calculations on full-size HTS cable concepts.

Table 7.1: Summary of coil, conductor and winding pack properties and requirements of the ITER and EU DEMO toroidal field coils.

conductor cross-section area	DEMO conductor
cable space	968.6 mm^2
copper stabilisation	554.9 mm^2
silver (HTS)	5.5 mm^2
helium void	272.2 mm^2
stainless steel in conduit	1056.2 mm^2
entire stainless steel	1147.6 mm^2
insulation	300 mm^2
operating current $I_{op}(13.27\ T, 5.0\ K)$	50 kA
critical current $I_c(13.27\ T, 5.0\ K)$	71.4 kA
safety factor (I_{op}/I_c)	0.7

Electrical stabilisation

The electrical stabilisation inside the cable is depending on the amount of additional copper. For DEMO TF coil conductors, the required copper cross-section are is 554.9 mm^2. Since the copper coating of *RE*BCO tapes does not cover the high copper ratio required in DEMO TF coils, additional copper is mandatory. This additional copper can either be plated, laminated or soldered around the *RE*BCO tapes, as it was done for the RACC to stabilise it electromechanically. Another method is to solder the superconductors into a cylindrical copper matrix, as it was done for most of the investigated TSTCs. The optimum solution is a combination of both: a thick copper stabilisation optimises the current sharing between superconductor and copper and transfers the current from the superconducting *RE*BCO layer to the surrounding rotationally symmetric copper matrix in case of a quench. Since all sample measurements were executed with technical high temperature superconductors, the thickness of the electrical *RE*BCO stabilisation will be taken for following scale-up calculations, as it is commercially available.

Mechanical stabilisation

With a cross-section area of 1147.6 mm^2, the mechanical reinforcement takes about 50% of the entirely available cross-section area. As seen in chapter 3.1.4, the stainless steel, respectively Hastelloy©substrate highly contributes to the internal stress-strain stability and therefore allows for a reduction of the stainless steel conduit. According to the executed FBI in-field strain measurements, the substrate thickness in following scale-up calculations is considered as it is commercially available for modern technical high temperature superconductors.

Flux jumping

Flux jumping denominates the effect of moving flux lines inside superconductors as a result of an induced or external changing magnetic field. The moving flux lines break cooper pairs inside the superconductor, which dissipates ohmic heat. This process can easily trigger a chain reaction, which would lead to an entire magnet quench. Hence, all conductors used for superconducting magnets have to be stable against flux jumps.

For low temperature superconductors, flux jump stability can be provided by using very thin superconductor filaments. The thinner the filaments are, the lower is the probability of flux jump propagation and the higher are the chances of cooling back to operation temperature. For 2G high temperature superconductors, the situation is different. As *RE*BCO superconductors are deposited on a substrate tape, flux jumps cannot be prevented by extruding the conductor into filigree filaments. However, in comparison to LTS, *RE*BCO benefits of better superconducting properties, such as a higher current sharing temperature and a higher critical current density. As we will see in the following, *RE*BCO does not necessarily require a filament structure to avoid flux jumps.

When a minimum of heat ΔQ_{ext} is supplied to a superconductor, the critical current density drops. Hitherto, superconducting screening currents become resistive currents, dissipating additional heat ΔQ_{int}. If external and dissipated heat is not covered by the effective volume specific heat capacity

$$\gamma C_{\text{eff}} = \frac{\Delta Q_{\text{ext}} + \Delta Q_{\text{int}}}{\Delta T}$$

of the surrounding material, a chain reaction of continuously dissipated screening currents is triggered. The flux lines jump and the magnetic field penetrates further into the superconductor. Figure 7.1 demonstrates the magnetic field before and after a flux jump.

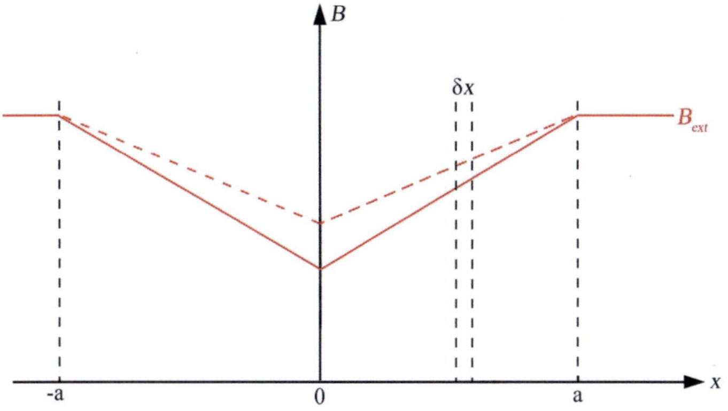

Figure 7.1: External magnetic field penetrating a superconductor before (continuous line) and after the flux jump (dashed line).

Under the assumption that an infinitesimal width δx encapsulates an infinitesimal magnetic flux change $\delta\phi(x)$, for the internal heat ΔQ accounts

$$\Delta Q_{\text{int}} = \int_a^b \delta q(x)\,\mathrm{d}x.$$

Since the electric energy is dissipated in heat, the heat change $\delta q(x)$ can be calculated by integrating over the resistive power

$$\delta q(x) = \int I(x)E(x)\,\mathrm{dt} = j_{\text{c}}\,\delta x\,\delta\phi(x).$$

Following this approach under consideration of the conductor geometry, its material properties and helium cooling, it can be shown, that the stability criterion of a superconducting slab against flux jumps can be described as

$$\beta_{\mathrm{lim}} = \frac{3}{i+3i^2}\left(1+\frac{4ha(1\text{-}\lambda)}{\pi^2\rho\gamma C}\mu_0(1+i)^2\right) \geq \frac{\mu_0\lambda^2 j_c^2 a^2}{\gamma C(T_C - T_{\mathrm{op}})} = \beta_{\mathrm{op}}, \tag{7.1}$$

where β, the stability parameter, defines the limit, at which triggers a flux jump. Mentioned variables are explained in table 7.2 and the derivation of equation 7.1 is attached in annex A.3. A more detailed derivation can be found in [Wil83].

Table 7.2: Variables and values to solve equation 7.1 for a Nb_3Sn tape, Nb_3Sn filament and a *RE*BCO tape.

variable	description	Cu stabilised Nb_3Sn tape	Cu stabilised Nb_3Sn filament	*RE*BCO tape
a	half conductor width (mm)	2	0.1	2
γC	volume specific heat capacity ($J \cdot m^{-3} \cdot K^{-1}$)	1000	1000	887
h	heat transfer coefficient ($W \cdot m^{-2} \cdot K^{-1}$)	50000	50000	50000
i	ratio of transfer current to critical current	0 - 1	0 - 1	0 - 1
$j_c(B = 13\ T)$	current density ($A \cdot m^{-2}$)	2.45×10^9	2.45×10^9	26×10^9
λ	filling factor	0.33	0.33	0.018
ρ	electrical resistivity of copper (Ωm)	2×10^{-10}	2×10^{-10}	2×10^{-10}
T_{op}	operation temperature	4.2 K	4.2 K	4.2 K
T_c	critical temperature	18.2 K	18.2 K	91 K

The left term of the equation 7.1 defines the stability parameter limit, depending on the ratio of transfer current to critical current inside the corresponding superconductor. The right term depends on the temperature margin between critical and operation temperature. In order to compare the flux jump stability of LTS with HTS, the stability parameter was calculated for a 0.2 mm wide Nb_3Sn filament and a 4 mm wide *RE*BCO tape. The results are shown in figure 7.2.

Both, the *RE*BCO tape and the Nb_3Sn filament almost share the same stability parameter, although the *RE*BCO conductor is 20 times wider. For Nb_3Sn with a 4 mm width, the stability factor would drop from 53 to 3. While filaments are mandatory for Nb_3Sn wire, *RE*BCO conductors do not require filaments to be stable against flux jumps. Especially, because this calculation only considers the heat capacity of the *RE*BCO surrounding copper stabilisation and totally neglects the stainless steel substrate. Taking the heat capacity of the stainless steel

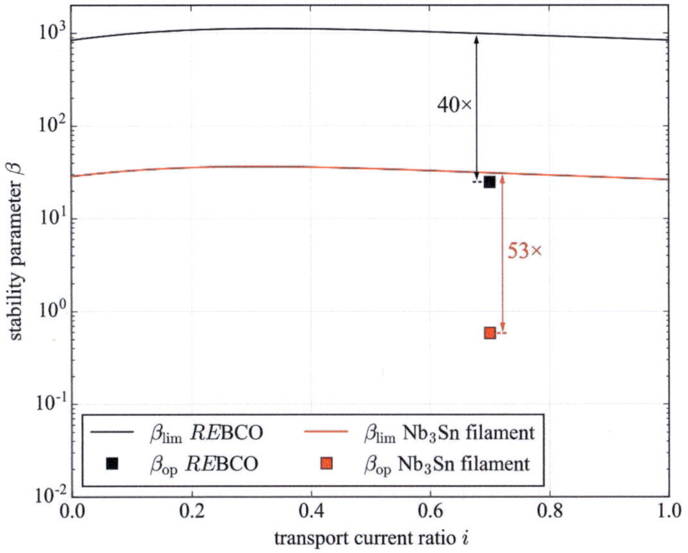

Figure 7.2: Stability parameters of a 0.2 mm wide Nb_3Sn filament and a 4 mm wide *RE*BCO tape depending on the ratio of transport current to critical current. For DEMO, the point of operation is at $i = 0.7$.

substrate into account increases the stability parameter even further. Flux jumping experiments for LTS have shown, that this calculation is very conservative and realistic scenarios exceed the expectations [Wil83].

Cooling

The forced flow cooling of the superconducting cable is provided by liquid helium pressed through the cable with sufficient helium void. Compared to the in-field performance and electromechanical stability, the cooling is a minor issue which can be optimised after a final cable-concept is selected. In order to already establish similar conditions as in the ITER CICC, following cable designs consider a helium void of minimal 272 mm^2, as it was estimated in chapter 3.3. For any of the following designs, the helium void is not crossed by conductor strands as it is the case for the ITER CICC. Therefore, the inlet pressure of an HTS DEMO TF coil is expected to be close or even lower than for an LTS ITER TF coil. Calculations and optimizations on the helium cooling was not part of this work.

Insulation

The required insulation of 300 mm^2 is a fix input parameter, which derives from calculations executed by [Gad16]. As it surrounds an entire coil winding and hence the entire full-size cable, it is not a parameter to be considered in the full-size cable design.

*RE*BCO conductor

For all cable designs investigated in chapter 6, it was shown, that as long as the cable concept provides sufficient electromechanical stability, it is the type of *RE*BCO tape which has the largest impact on the cable performance. Except for sample C, all tested cable designs proved to withstand Lorentz forces which corresponds to a tape pressure p_{tape} of approximately 20 MPa. In order to establish fair comparison during the scale-up of the different cable designs, the scale-up is calculated for commercially available[1] SuperPower SCS4050-AP *RE*BCO tapes, with an ordinary in-field performance and a thick copper stabilisation. The exact data of the most adequate HTS wire is listed in table 7.3.

Table 7.3: *RE*BCO conductor used for up-scaling calculations.

parameter	values
*RE*BCO wire	SuperPower SCS4050-AP
tape width	4 mm
tape I_c(77 K)	150 A
cable I_c(4.2 K, 13 T)	210 A
substrate	50 μm
copper stabilisation	110 μm
silver	2 μm
total thickness	165 μm

7.2 Full-Size Stacked Tape Cable (FS-STC)

The stacked tape cable is a straight-forward approach to a full-size TF coil cable. FBI measurements in chapter 6 have shown, that the TSTC with soldered tape stacks is a reliable design, which never showed unexpected degradation due to temperature dependent in-field measurements or longitudinal tensile stress and n-values between 30 and 40.

[1] According to the SuperPower inventory list of July 2016.

7.2.1 Up-scaling

Considering the single tape parameters from table 7.3, the required entire superconductor width is

$$w_{\text{tot}} = \frac{I_{\text{c-DEMO}}}{I_{\text{c-tape}}} \cdot w_{\text{tape}} = \frac{71.4\,\text{kA}}{210\,\text{A}} = 1361\,\text{mm}.$$

This corresponds to 342 *RE*BCO tapes with a width of 4 mm. Divided into three stacks. Each stack contains 19 tapes with an entire stack height of 3.14 mm. As the simple tape stacks do not provide sufficient copper stabilisation for a full-size DEMO conductor, they need to be embedded inside a copper matrix. A very suitable solution is a Rutherford winding around a rectangular copper former. The angle of the helical winding is 83.3° and its slope is tan(83.3°) = 8.51. With a former radius of 7.28 mm, this corresponds to an arc length of 196.0 mm with a curvature radius of 534.8 mm. This curvature radius was selected to satisfy the strain conditions, which will be discussed in subsection 7.4.2.

In order to electromechanically stabilise the superconductors the 18 stacks are embedded inside a copper matrix. Together with the copper of the *RE*BCO tapes, the entire copper cross-section area is

$$\begin{aligned} A_{\text{Cu}} &= w_{\text{tape}} \cdot n_{\text{tapes}} \cdot A_{\text{Cu per mm tape}} + A_{\text{Cu-matrix}} \\ &= 4\,\text{mm} \cdot 342 \cdot 0.11\,\text{mm}^2/\text{mm} + 448.3\,\text{mm}^2 = 598.8\,\text{mm}^2, \end{aligned}$$

which is 8.2% more copper than required.

For stainless steel, the cross-section area consists of the steel conduit and the substrate of the *RE*BCO tapes, which is

$$\begin{aligned} A_{\text{ss}} &= w_{\text{tape}} \cdot n_{\text{tapes}} \cdot A_{\text{ss per mm tape}} + A_{\text{ss-conduit}} \\ &= 4\,\text{mm} \cdot 342 \cdot 0.05\,\text{mm}^2/\text{mm} + 1077.8\,\text{mm}^2 = 1146.2\,\text{mm}^2, \end{aligned}$$

which almost precisely fits the DEMO requirement.

The helium void is considered with 272.2 mm^2, distributed to 16 helium channels with 210.0 mm^2 and the helium void of 62.2 mm^2 between the cable and the conduit.

A schematic of the FS-STC, which satisfies the current and spatial requirements, is shown in figure 7.3. A comparison of the DEMO and the FS-STC properties is listed in table 7.5 in section 7.6. Except for the silver ratio, which is an arbitrary parameter, all values are met or even exceeded.

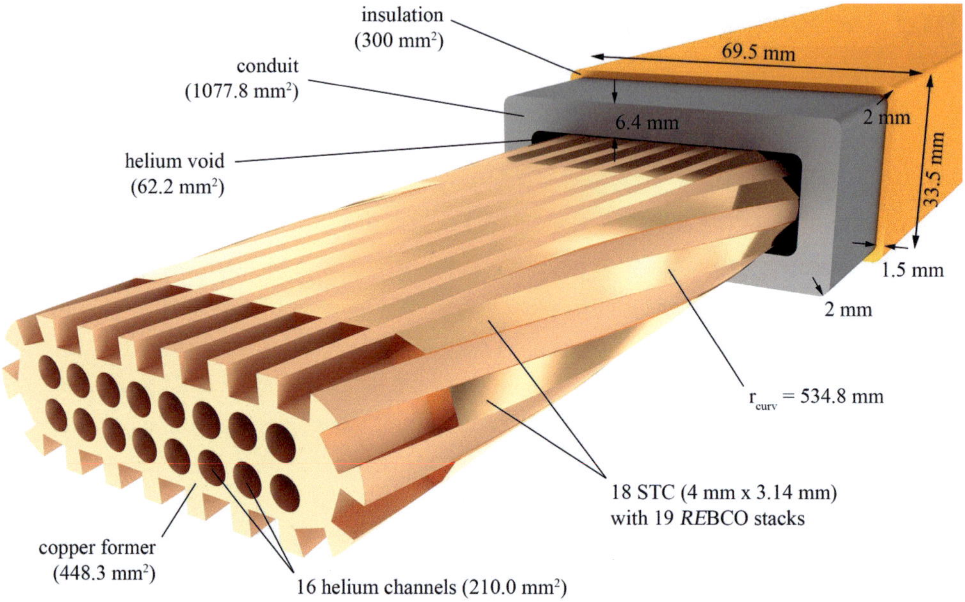

Figure 7.3: Schematic of a feasible DEMO full-size STC conductor.

7.2.2 Longitudinal strain

Due to the Rutherford winding, the tape stacks experience tensile strain, which derives from the helical winding around the Rutherford former and the stack twist, obligatory to keep the stack plane parallel to the groove surface. Under the assumption, that both, the twist pitch h and the curvature radius R_{curv} of the helix exceed the tape width w and the stack thickness t by far (h, $R_{\mathrm{curv}} \gg w, t$), the experienced strain can be calculated by

$$\varepsilon = \frac{1}{2}\left(\frac{2\pi}{h}\right)^2\left(x^2 - \frac{y^2}{12}\right) + \frac{x\cos\left(\frac{2\pi}{h}s\right) - y\sin\left(\frac{2\pi}{h}s\right)}{R_{\mathrm{curv}}}, \tag{7.2}$$

with the coordinates $-\frac{w}{2} \leq x \leq \frac{w}{2}$ and $-\frac{t}{2} \leq y \leq \frac{t}{2}$ [BUWB15].

As the twist pitch of the stacks is equivalent to the pitch of the helix, the stack twist only contributes with less than 15% to the entire strain. The maximum strain on the innermost tape 1 was set to 0.5%. This corresponds to a curvature radius of 438.7 mm and therefore defines the helix winding angle with 83.4° and the pitch of a stack half-twist with 157.4 mm for the innermost tape. With increasing tape layer, the curvature radius increases and the experienced strain decreases. Together with a longitudinal strain of < 0.1% (3.3) due to coil hoop stress, the targeted maximum strain level of 0.6%, which was measured in chapter 6, section 6.2.5, is satisfied for all tape layers. The calculation was executed for seven tapes of the stack over their entire twist pitch, including the innermost tape 1 and the outermost tape 19. Results are shown in 7.4.

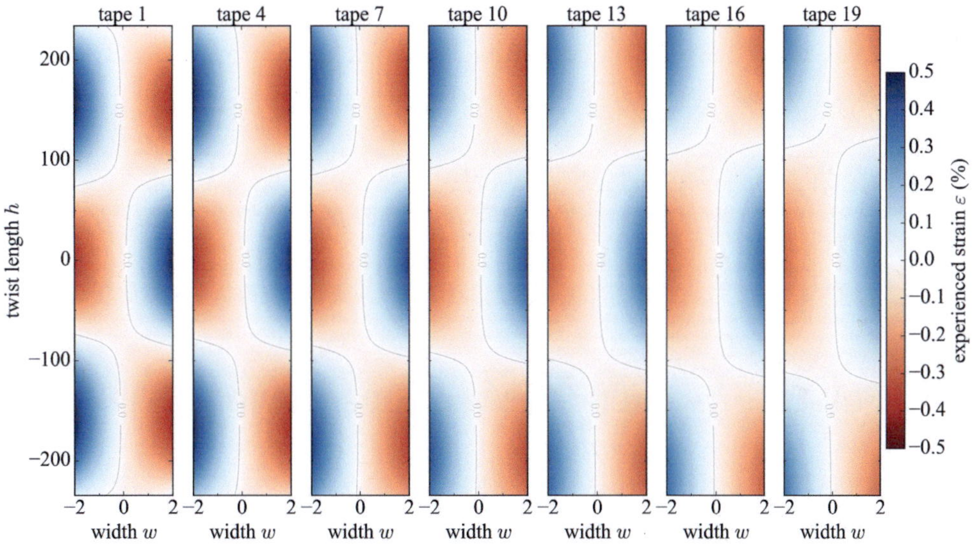

Figure 7.4: Strain calculation for seven of 19 tapes inside the 18 twisted and bent *RE*BCO stacks. The peak strain in the innermost tape 1 is 0.5%.

7.2.3 Contacts

As already discussed in chapter 4, contacts to extend the cable length should only be executed on tape level. With actual single piece lengths of about 500 m and a number of 19 tapes per stack, this corresponds to a lap joint every 26 m. Since each tape can be extended individually, a face-to-face lap-joint is the optimum solution. Spatial limitations for the overlap length are not given and under consideration of all parameters discussed in chapter 4, joint resistances of $R_j < 25\,\mathrm{n\Omega}$ at 4.2 K per lap joint can be easily achieved. With an operation current of 210 A per tape, the heat input per joint is only $P_j = 1.1\,\mathrm{mW}$. For one entire coil the heat input, depending on the coil perimeter p_{coil}, the number of coil windings n_{wind}, the number of tapes per winding $n_{tapes}$and the single piece length l_{sp}, totals in

$$\begin{aligned} P_{j-TF} &= p_{coil} \cdot n_{wind} \cdot n_{tapes}/l_{sp} \cdot P_j \\ &= 39.83\,\mathrm{m} \cdot 384 \cdot 342/500\,\mathrm{m} \cdot 1.1\,\mathrm{mW} = 11.51\,\mathrm{W}. \end{aligned}$$

7.2.4 Conductor length

The length of 4 mm wide *RE*BCO superconductors, required for one entire TF coil, can be roughly estimated by multiplying the coil perimeter p_{coil} with the number of coil windings n_{wind}, the ratio of the helix pitch to its arc length c_{helix} and the entirely required tape width w_{tot} per winding, divided by the single tape width w_{tape}, which is

$$\begin{aligned} l_{RE\text{BCO}} &= p_{\text{coil}} \cdot n_{\text{wind}} \cdot c_{\text{helix}} \cdot w_{\text{tot}}/w_{\text{tape}} \\ &= 39.83\,\text{m} \cdot 384 \cdot 1.007 \cdot 1368\,\text{mm}/4\,\text{mm} = 5,267,409\,\text{m}. \end{aligned}$$

For 16 TF coils, the total conductor length of 4 mm *RE*BCO tape is approximately 84,300 km.

7.3 Full-Size Twisted Stacked Tape Cable (FS-TSTC)

Assuming that future experiments reveal that tape twist is a required feature to successfully ramp and operate the TF magnets, the Full-Size Twisted Stacked Tape Cable (FS-TSTC) is a very promising approach. Various *RE*BCO tapes of low width are twisted and embedded into a copper matrix and afterwards wound around a central copper former. Measurements in chapter 6 have shown, that the FS-TSTC with soldered and twisted tape stacks is a reliable design, which never showed degradation due to temperature dependent in-field measurements. High n-values between 30 and 40 demonstrate a sharp superconducting transition. As no sample was damaged at currents beyond the critical current criterion, the TSTC proved to be a cable with high inherent, electromechanical stability.

7.3.1 Stack variants

Two different TSTC variants were investigated in chapter 6. The first variant was a *RE*BCO stack with tapes of the same width (sample STC A and TSTC E) and the second variant was made of tapes with two different widths (sample STC B, TSTC D and TSTC F), forming a cross-shaped tape stack. To satisfy the entire width of superconductor 1361 mm are required. This can either be distributed in variant 1 configuration or variant 2 configuration. Variant 1 consists of 18 superstrands with nineteen 4 mm wide tapes each and variant 2 consists of 18 superstrands with fifteen 4 mm wide tapes and eight 2 mm wide tapes each. Both strand variants are demonstrated in figure 7.5. As both variants are compatible, the following upscaling will be calculated for both of them. With exactly the same amount of *RE*BCO conductor, only the strain calculation will show different results.

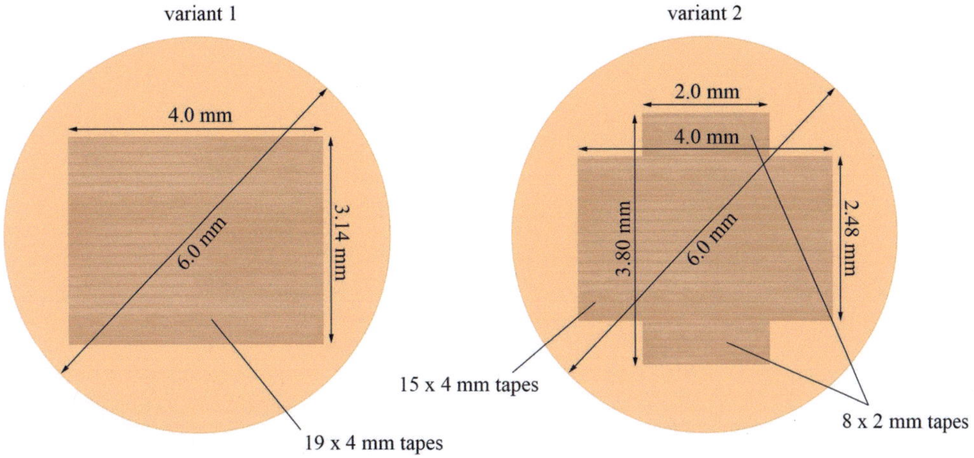

Figure 7.5: Comparison of both TSTC variants. The amount of conductor and the copper ratio is the same.

7.3.2 Up-scaling

Each stack is twisted around the axis in longitudinal direction with a twist pitch of 320 mm. This twist pitch corresponds to the twist pitch of the TSTC sample E investigated in chapter 6. Afterwards each stack is embedded into a copper matrix with an outer diameter of 6.0 mm. Each of the 18 resulting superstrands is helically wound around a 12 mm thick rectangular copper former. The helix angle is 84.0° and its slope is tan(84°) = 9.52. For each of the half-windings on the short former sides, this corresponds to an arc length of 180.5 mm with a curvature radius of 550.1 mm. This parameters were selected for reasons discussed in subsection 7.3.3.

This cable geometry is only feasible with a stainless steel conduit thickness of 6.25 mm, which corresponds to a stainless steel cross-section of

$$\begin{aligned} A_{\mathrm{ss}} &= n_{\mathrm{tapes}} \cdot w_{\mathrm{tape}} \cdot A_{\mathrm{ss\,per\,mm\,tape}} + A_{\mathrm{ss-conduit}} \\ &= 342 \cdot 4\,\mathrm{mm} \cdot 0.05\,\mathrm{mm}^2/\mathrm{mm} + 1056.25\,\mathrm{mm}^2 = 1124.7\,\mathrm{mm}^2, \end{aligned}$$

and therefore lags 22.9 mm^2 for the DEMO requirement[2]. To compensate the missing cross-section area, two 36 mm wide stainless steel tapes are embedded inside the copper matrix to increase the entire stainless steel cross-section to 1147.6 mm^2. The helium void of 272.8 mm^2 is distributed among the central channel (119.2 mm^2) and the void around the superstrands (153.6 mm^2). The remaining cross-section area is filled with copper. This includes the copper

[2] As the orientation of the twisted stacks inside the cable is arbitrary, cross-section areas are calculated on basis of the default tape width

stabilisation of the superconductors, the copper strands and the central former and totals in

$$\begin{aligned} A_{\mathrm{Cu}} &= n_{\mathrm{tapes}} \cdot w_{\mathrm{tape}} \cdot A_{\mathrm{Cu\,per\,mm\,tape}} + A_{\mathrm{Cu-strands}} + A_{\mathrm{Cu-former}} \\ &= 342 \cdot 4\,\mathrm{mm} \cdot 0.11\,\mathrm{mm}^2/\mathrm{mm} + 283.14\,\mathrm{mm}^2 + 159.1\,\mathrm{mm}^2 = 592.7\,\mathrm{mm}^2. \end{aligned}$$

With these numbers given, the FS-TSTC fully satisfies the DEMO requirements. The summarizing comparison is found in table 7.5 in section 7.6. A schematic view of the cable design is given in figure 7.6.

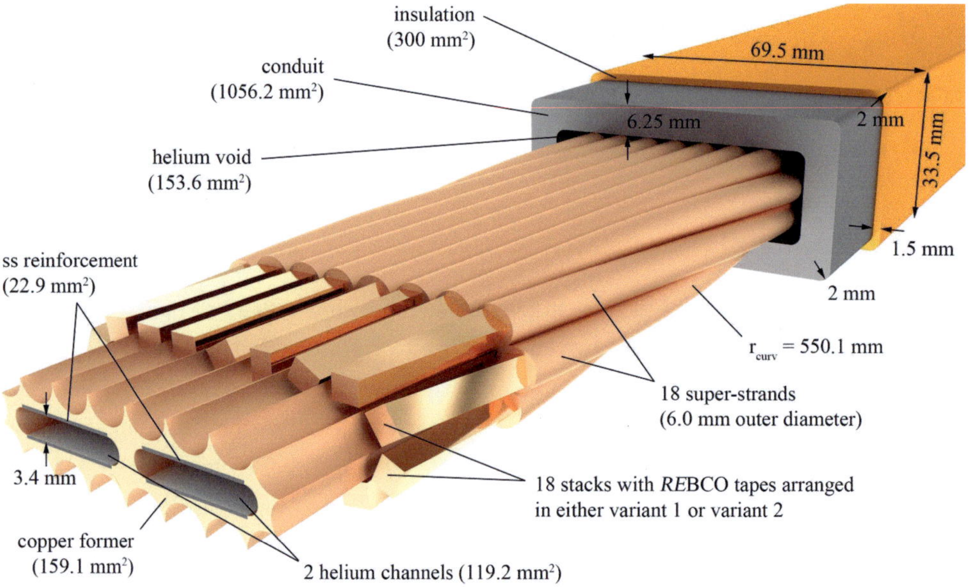

Figure 7.6: Schematic of a feasible DEMO full-size TSTC conductor.

7.3.3 Longitudinal strain

Longitudinal strain is the limiting factor considering the FS-TSTC design. Both the twisting and the winding of the superstrands around its narrow central copper former elongates the superconductor tapes and leads to tensile strain. Under the assumption, that both, the twist pitch h and the curvature radius R_{curv} of the helix exceed the tape width w and the stack thickness t by far (h, $R_{\mathrm{curv}} \gg w$, t), the experienced strain can be calculated by equation 7.2.

The twist pitch of 320 mm was taken as fix parameter, as it was adapted from TSTCs investigated in chapter 6, which have shown very good results for in-field and temperature dependent measurements. In order to limit the experienced strain at the outermost tapes of both variants to 0.5%, the minimum curvature radius R_{curv} is set to 550.0 mm. Combined with the expected longitudinal strain of < 0.1% due to coil hoop stress during operation, the upper strain limit of 0.6%, which was measured in chapter 6.2.5, is satisfied. The strain experienced in the *RE*BCO stacks is visualised in figure 7.7.

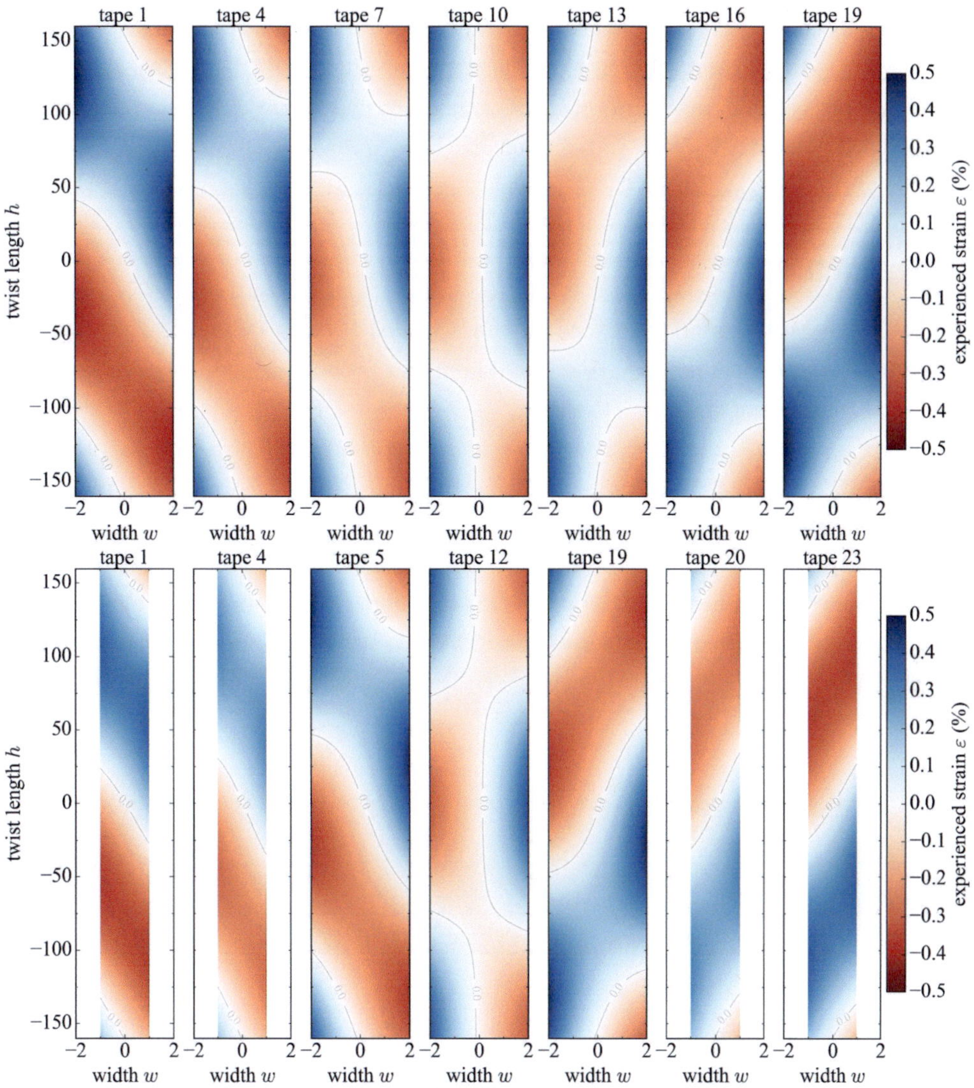

Figure 7.7: Strain calculation for variant 1 and variant 2 (top down) of the tapes inside the 18 twisted and bent *RE*BCO stacks. For variant 1 the peak strain is 0.50% in tape 1 and tape 19. For variant 2 the peak strain is 0.47% in tape 5 and tape 19.

For variant 1 the outermost tapes are tape 1 and tape 19 and the centre tape is tape 10. For variant 2, the outermost 2 mm wide tapes are tape 1 and tape 23, the innermost 2 mm tapes are tape 4 and tape 20, the outermost 4 mm tape is tape 5 and tape 19, and the centre tape is tape 12. The calculation was executed for the entire twist length of 320 mm. As expected, the innermost tapes experience very low tensile strain. The highest tensile strain was measured for tape 1 and tape 19 of variant 1 with a maximum strain of 0.5%. For variant 2, the peak strain is 0.47% at tape 5 and tape 19. According to foregoing investigations, strain at 0.5% has no negative effect on the tape respectively cable performance as long as the upper limit of 0.6% is not exceeded. On basis of this strain calculation with the outcome of a required curvature radius of 550 mm at a twist pitch of 320 mm, the FS-TSTC concept was designed.

7.3.4 Contacts

The optimum feasible solution are face-to-face lap joints of tapes with a single piece length of 500 m. The heat input per joint $P_j = 1.1\,\text{mW}$ remains for both variants the same as for the FS-STC. The total heat input per TF coil is

$$\begin{aligned} P_{\text{j-TF}} &= p_{\text{coil}} \cdot n_{\text{wind}} \cdot n_{\text{tapes}} / l_{\text{sp}} \cdot P_j \\ &= 39.83\,\text{m} \cdot 384 \cdot 342/500\,\text{m} \cdot 1.1\,\text{mW} = 11.51\,\text{W}. \end{aligned}$$

Due to the curvature radius of 550.1 mm and the twist pitch of 320 mm, the tapes and the lap joints experience tensile strain values of up to 0.5%. However, as discussed in chapter 4, the tensile strain does not impair the joint quality. The weak spot is always the *RE*BCO tapes themselves.

7.3.5 Conductor length

For both variants the conductor length is the same. The very large helix angle of 84° does not increase the amount of required tapes significantly. The entire length has only to be corrected by c_{helix}, which is the ratio of the helix pitch to its arc length. For one single TF coil the conductor length of 4 mm wide *RE*BCO is

$$\begin{aligned} l_{RE\text{BCO}} &= p_{\text{coil}} \cdot n_{\text{wind}} \cdot c_{\text{helix}} \cdot w_{\text{tot}} / w_{\text{tape}} \\ &= 39.83\,\text{m} \cdot 384 \cdot 1.006 \cdot 1368\,\text{mm}/4\,\text{mm} = 5,262,179\,\text{m}. \end{aligned}$$

For 16 TF coils, the total conductor length is about 84,200 km.

7.4 Full-size RACC (FS-RACC)

The RACC cable is the only HTS cable design with full tape transposition. In applications with fast magnet ramps and high amounts of AC current losses, the RACC dominates over any other cable design. Therefore, the RACC is recommended for magnets of particle accelerators and superconducting generators. During the hitherto manufacturing process a RACC requires twice as much *RE*BCO material as a STC or TSTC. Furthermore, as seen in chapter 6.1, its mechanical resilience against Lorentz forces is very low and demands for sophisticated reinforcement.

7.4.1 Up-scaling

As the RACCs do not provide sufficient copper stabilisation for a full-size DEMO conductor, they need to be embedded inside a copper matrix. In order to satisfy the DEMO requirement of 1361 mm superconductor width, 18 RACCs with 38 2 mm wide *RE*BCO Roebel strands each, are required. The height of each RACC is approximately 3.14 mm and its transposition length is 262.4 mm. All RACCs are embedded inside the grooves of a 17.7 mm thick rectangular copper matrix. This entire composition forms the Roebel Rutherford Conductor. To guarantee sufficient electromechanical stability, the RACCs needs to be soldered inside the former. Otherwise, the current-sharing between the conductor and the copper matrix is insufficient in case of a cable quench.

The angle of the helical winding is 83.3° and its slope is tan(83.3°) = 8.51. With a former radius of 7.28 mm, this corresponds to an arc length of 196.0 mm with a curvature radius of 534.8 mm. This curvature radius is the minimal radius to satisfy the strain conditions, which will be discussed in subsection 7.4.2.

For the FS-RACC, the thickness of the stainless steel conduit can be kept at 6.4 mm. The total stainless steel cross-section area is

$$\begin{aligned} A_{\text{ss}} &= n_{\text{tapes}} \cdot w_{\text{tape}} \cdot A_{\text{ss per mm tape}} + A_{\text{ss-conduit}} \\ &= 684 \cdot 2\,\text{mm} \cdot 0.05\,\text{mm}^2/\text{mm} + 1077.76\,\text{mm}^2 = 1146.2\,\text{mm}^2, \end{aligned}$$

which satisfies the DEMO requirements.

The helium void of 272.2 mm^2 is mainly covered with 210.0 mm^2 by the 16 helium channels inside the copper matrix and the missing 62.2 mm^2 come from the void between cable and conduit. All remaining space is filled with the copper matrix.

The entire copper cross-section area, including the former and the *RE*BCO tapes, is

$$A_{\mathrm{Cu}} = w_{\mathrm{tape}} \cdot n_{\mathrm{tapes}} \cdot A_{\mathrm{Cu\,per\,mm\,tape}} + A_{\mathrm{Cu-former}}$$
$$= 684 \cdot 2\,\mathrm{mm} \cdot 0.11\,\mathrm{mm}^2/\mathrm{mm} + 448.3\,\mathrm{mm}^2 = 598.8\,\mathrm{mm}^2$$

and therefore slightly exceeds the DEMO requirement by 43.9 mm^2. A summary of the cable data and a comparison with all other cable designs is found in table 7.5 in section 7.6. A schematic view of the FS-RACC is given in figure 7.8.

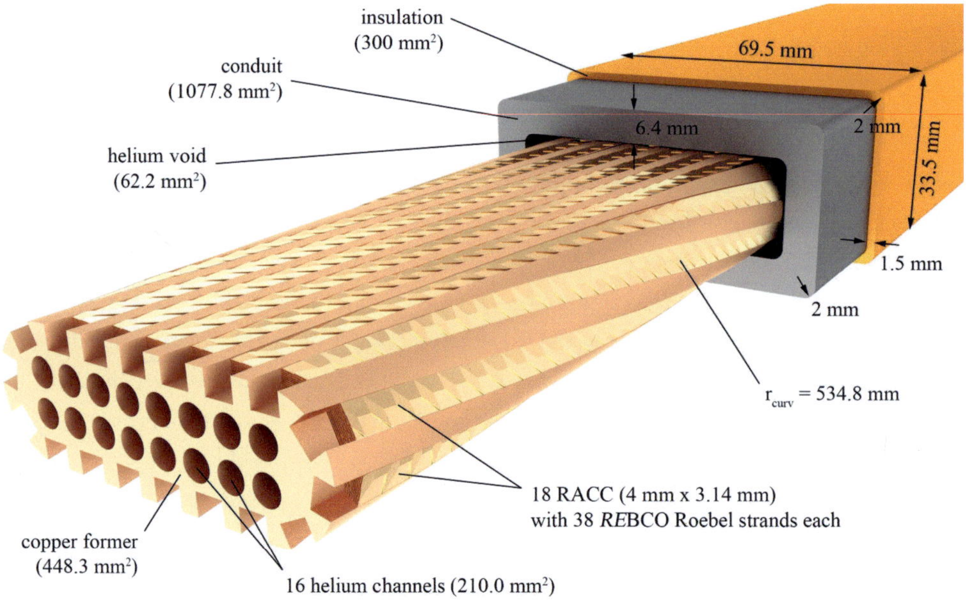

Figure 7.8: Schematic of a feasible DEMO full-size RACC cable.

7.4.2 Longitudinal strain

Due to the Rutherford winding, the RACC strands experience tensile strain. This strain derives from the twist of the RACCs, which is necessary to keep the RACC plane parallel to the groove surface. As the twist pitch of the RACCs is equivalent to the pitch of the helix, the RACC twist only contributes with less than 10% to the entire strain which again is calculated with equation 7.4. Adaptions due to the Roebel structure are not required. The impact of the tape transposition inside a RACC, is negligible compared to the curvature radius and therefore only contributes with less than 0.2% to the calculated strain.

As for all foregoing cable designs, the experienced strain on the outermost tapes was set to 0.5%. This corresponds to a curvature radius of 438.7 mm and therefore defines the helix winding angle with 83.4° and the pitch of a RACC half-twist with 157.4 mm for the innermost layer. Again, together with a longitudinal strain of < 0.1% due to coil hoop stress, the targeted maximum strain of 0.6% is satisfied for all tape layers. The results of the strain calculation are visualised in figure 7.9 for all layers. The calculation was explicitly executed for seven different layers along the twist pitch of the 4 mm wide RACC and not for particular Roebel strands. Due to their transposition, the Roebel strands shift within these layers and therefore it is not reasonable to focus on individual strands, but to consider the entire area they pass.

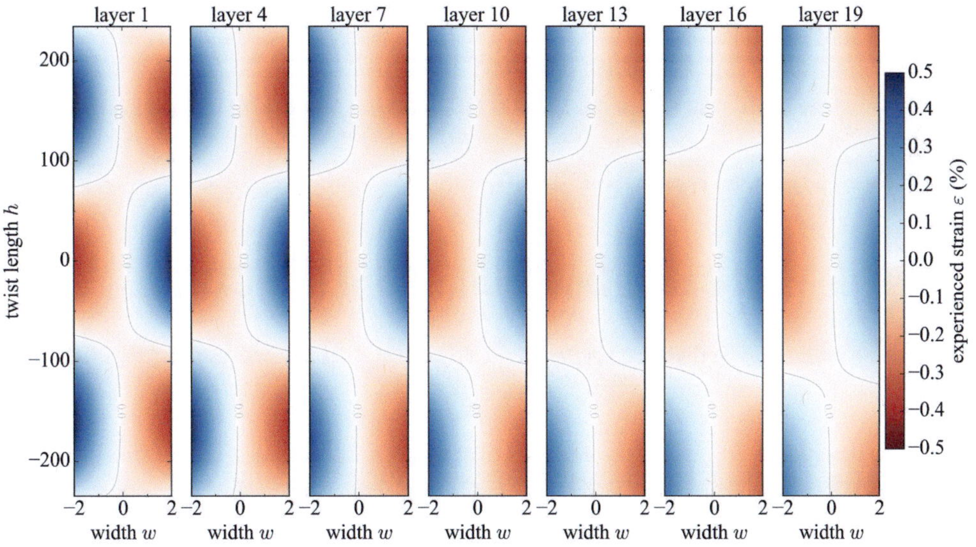

Figure 7.9: Strain calculation for seven of 19 layers inside the 18 twisted and bent RACC. The peak strain in the innermost layer 1 is 0.50%.

7.4.3 Contacts

As for both foregoing full-size conductors, the contacts to extend the FS-RACC length needs to be executed on tape level. For practical reasons, this extension needs to be executed before punching the Roebel strands.

The overall heat input for the FS-RACC remains almost the same as for the FS-TSTC. With an expected heat input per joint of $P_\mathrm{j} = 0.6\,\mathrm{mW}$, the total heat input per TF coil is

$$\begin{aligned} P_{\mathrm{j-TF}} &= p_{\mathrm{coil}} \cdot n_{\mathrm{wind}} \cdot n_{\mathrm{tapes}} / l_{\mathrm{sp}} \cdot P_\mathrm{j} \\ &= 39.83\,\mathrm{m} \cdot 384 \cdot 684 / 500\,\mathrm{m} \cdot 0.6\,\mathrm{mW} = 12.6\,\mathrm{W}. \end{aligned}$$

According to the discussion in chapter 4, joint damage due to tensile strain is not expected.

7.4.4 Conductor length

At actual point of development, a RACC strand requires twice the material as any tape stack. Punching a 2 mm wide strand out of a 4 mm wide strand produces almost 50% chad. This could be reduced by punching or laser-cutting the Roebel strands out of very wide *RE*BCO raw material. However, at the moment the maximum, commercially produced tape width is only 12 mm. In the optimum case, five strands could be cut from this tape width, which still requires 1.2 more *RE*BCO material than any simple tape stack. The helix angle of 83.3 ° does not increase the amount of required tapes significantly. With the chad-value c_{chad} and the correction value c_{helix}, the conductor length becomes

$$\begin{aligned} l_{RE\text{BCO}} &= p_{\text{coil}} \cdot n_{\text{wind}} \cdot c_{\text{helix}} \cdot c_{\text{chad}} \cdot w_{\text{tot}} / w_{\text{tape}} \\ &= 39.83\,\text{m} \cdot 384 \cdot 1.007 \cdot 1.2 \cdot 1368\,\text{mm}/4\,\text{mm} = 6,320,891\,\text{m}. \end{aligned}$$

For 16 TF coils, the total sum is about 101,100 km.

7.5 Full-size Cable On Round Core (FS-CORC)

Last cable candidate to be discussed is the FS-CORC. The CORC is well known for its flexibility and tensile strength. However, the conductors are only semi-transposed and the low current sharing and bad heat exchange between the layered *RE*BCO windings and the surrounding copper former is a serious issue, already addressed in chapter 6.

7.5.1 Up-scaling

The required conductor width of 1361 mm superconductor is distributed within 17 CORC strands. Each strand consists of eleven 3 mm wide and twelve 4 mm wide tapes, which are wound around a former with an outer diameter of 2.97 mm. The number of tapes per layer depends on the perimeter of the foregoing layer respectively former and the winding angle of 45°, which is the angle with the highest CORC tensile strength [LAP+11]. Each CORC strand consists of 10 wound *RE*BCO layers. The number of tapes per layer is listed in table 7.4. The resulting outer CORC diameter is 6.27 mm.

In order to electro-mechanically stabilise the CORC strands they are wound around an 11.71 mm thick copper former. The helical winding angle depends on the minimum bending

Table 7.4: Layer winding of a CORC strand.

Layer	1	2	3	4	5	6	7	8	9	10
perimeter (mm)	9.33	10.37	11.4	12.44	13.48	14.51	15.55	16.59	17.62	18.66
tape width (mm)	3	3	3	3	3	4	4	4	4	4
number of tapes	2	2	2	2	3	2	2	2	3	3

radius of a CORC cable, which is less than 3 cm [VDL+14]. Aiming for a 3 cm minimum bending radius, the helix angle was set to 81.9° with a slope of tan(81.9°) = 7.03. This corresponds to a curvature radius of 302 mm and an arc length of 133.78 mm for each half-winding along the short side of the copper former.

The stainless steel conduit thickness has to be kept at 6.25 mm. Otherwise it is not possible to provide enough space for the CORC strands. The entire stainless steel cross-section is

$$\begin{aligned} A_{\text{ss}} &= n_{\text{tapes}} \cdot w_{\text{tape}} \cdot A_{\text{ss per mm tape}} + A_{\text{ss-conduit}} \\ &= (187 \cdot 3\,\text{mm} + 204 \cdot 4\,\text{mm}) \cdot 0.05\,\text{mm}^2/\text{mm} + 1056.25\,\text{mm}^2 = 1125.1\,\text{mm}^2, \end{aligned}$$

which lags 22.5 mm^2 for the DEMO requirement. This missing cross-section area is compensated by two 36 mm wide stainless steel tapes, embedded inside the copper matrix.

The entire helium void of 272.8 mm^2 is distributed among the central channels (107.7 mm^2) and the void around the superstrands (165.1 mm^2).

For the required copper stabilisation, including *RE*BCO material, the CORC formers and the central former, accounts

$$\begin{aligned} A_{\text{Cu}} &= w_{\text{tape}} \cdot n_{\text{tapes}} \cdot A_{\text{Cu per mm tape}} + A_{\text{CORC-formers}} + A_{\text{Cu-former}} \\ &= (187 \cdot 3\,\text{mm} + 204 \cdot 4\,\text{mm}) \cdot 0.11\,\text{mm}^2/\text{mm} + 117.8\,\text{mm}^2 + 145.5\,\text{mm}^2 = 414.8\,\text{mm}^2. \end{aligned}$$

The FS-CORC does not satisfy the DEMO copper requirements. Due to void between the wound *RE*BCO tapes it lags 140.1 mm^2. The CORC parameters are compared with the parameters of foregoing cable designs in table 7.5, section 7.6. A schematic view of the cable design is given in figure 7.10.

7.5.2 Contacts

Contacts to extend the FS-CORC should be executed on tape level. The overall heat input for the FS-CORC remains the same as for the FS-TSTC, corrected by the extended conductor length of $\sqrt{2}$ due to the 45° winding angle.

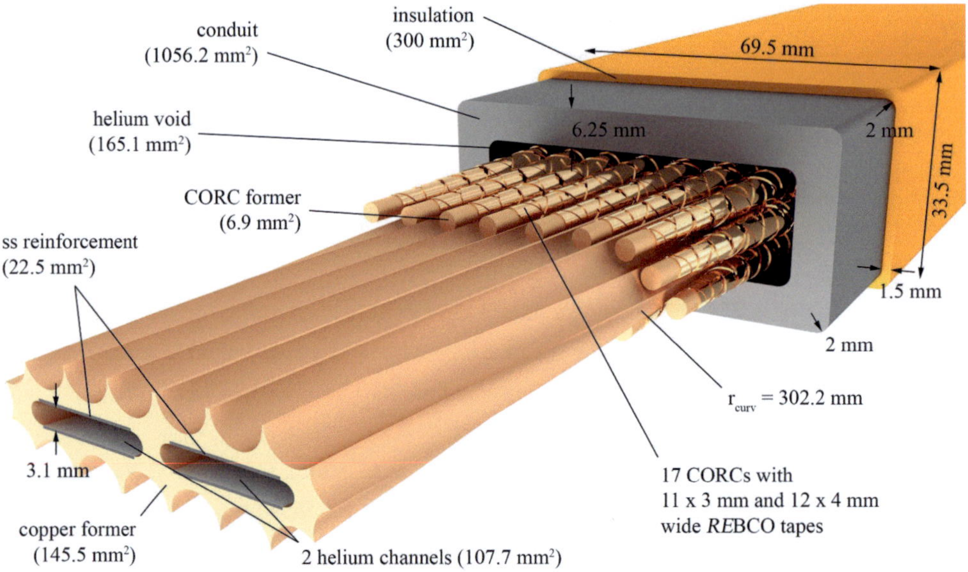

Figure 7.10: Schematic of a feasible DEMO full-size CORC cable.

With an expected heat input per joint of $P_\mathrm{j} = 1.1\,\mathrm{mW}$, the total heat input per TF coil is

$$\begin{aligned} P_\mathrm{j-TF} &= p_\mathrm{coil} \cdot n_\mathrm{wind}/l_\mathrm{sp} \cdot n_\mathrm{tapes} \cdot P_\mathrm{j} \cdot \sqrt{2} \\ &= 39.83\,\mathrm{m} \cdot 384/500\,\mathrm{m} \cdot (204 \cdot 1.1\,\mathrm{mW} + 187 \cdot 1.1\,\mathrm{mW} \cdot \frac{3}{4}) \cdot \sqrt{2} = 16.4\,\mathrm{W}. \end{aligned}$$

According to the discussion in chapter 4, joint damage due to tensile strain is not expected.

7.5.3 Conductor length

The 45° winding angle of each CORC strand increases the required *RE*BCO material by a factor of $\sqrt{2}$. With the helix angle of 81.9° the entire length has to be corrected by c_helix, which is the ratio of the helix pitch to its arc length. The conductor length of 4 mm wide *RE*BCO tape for one TF coil is

$$\begin{aligned} l_{RE\mathrm{BCO}} &= p_\mathrm{coil} \cdot n_\mathrm{wind} \cdot c_\mathrm{helix} \cdot w_\mathrm{tot}/w_\mathrm{tape} \cdot \sqrt{2} \\ &= 39.83\,\mathrm{m} \cdot 384 \cdot 1.01 \cdot 1377\,\mathrm{mm}/4\,\mathrm{mm} \cdot \sqrt{2} = 7{,}520{,}589\,\mathrm{m}. \end{aligned}$$

For 16 TF coils, the total conductor length is about 120,300 km.

7.6 Summary and evaluation

Three full-size fusion cable concepts satisfy the DEMO TF magnet coil requirements. Among these three cables is one full-size Stacked Tape Cable (FS-STC), one full-size Twisted Stacked Tape Cable (FS-TSTC) and one full-size Roebel Assembled Coated Conductor (FS-RACC). The exact parameters for each cable and the comparison with the DEMO requirements are listed in table 7.5.

The first two cable designs, the STC and the TSTC, were calculated on the basis, that full tape transposition is considered a minor issue for 2G HTS TF coils. Firstly, HTS conductors are very stable against flux jumps and secondly, the heat input due to eddy currents can be reduced by decreasing the magnet ramp rate, which is freely adjustable for the TF coils.

The FS-STC is the most straight forward approach to a full-size cable. It is easy to manufacture, does not require a twist along the tape stack and is well embedded inside the copper former which provides for a continuous heat exchange and current sharing between the superconductors and the copper stabilisation. It meets all DEMO requirements and even exceeds the copper ratio by 43 mm^2.

The FS-TSTC is another very promising stacked tape approach with an additional stack twist around its centre line. Two variants have been proposed. Variant 1 is easier to manufacture, while variant 2 experiences 0.3% less strain during the winding process. For the FS-TSTC heat exchange and current sharing is provided by the *RE*BCO tapes surrounding copper tubes. During many in-field measurements not one single STC or TSTC that was soldered into a copper tube, was destroyed or damaged by fusing. Even during cable quench, the very good current sharing with the copper matrix always protected the samples from fusing and left sufficient time to trigger the interlock. This makes the FS-STC respectively the FS-TSTC an optimum candidate for fusion applications, where electrical stability is of highest priority.

A RACC benefits from his full tape transposition and low AC losses and is therefore the optimum candidate for accelerator- or generator magnets. In-field measurements have shown, that a RACC is more difficult to electromechanically stabilise than any other cable design. This issue has been addressed and extensively discussed in chapter 6.

Among all full-size cable designs, the FS-CORC is the only cable that does not satisfy the required copper cross-section area. One approach to increase the copper ratio is to reduce the helium void inside the copper former. This approach is satisfied by the fact, that void inside the CORC strands was not considered. It was neglected for the fact, that if included in the overall void, a higher helium pressure is necessary to achieve the same helium flow rates. However, even apart from the copper ratio, the FS-CORC does not qualify as TF coil cable. The main problem of the FS-CORC is the current sharing with the copper former. The CORC windings are wound with the *RE*BCO substrate to the outside and each layer is wound in opposite direction.

This reduces the interlinks between each layer significantly and for each layer, the current has to bypass the stainless steel substrate of the *RE*BCO tapes. Only the outermost tapes are in direct contact with the copper former. The main qualities of a CORC are its flexibility and its tensile strength, which make the CORC great a cable for power transmission. However, in a DEMO TF coil neither flexibility nor tensile strength are mandatory requirements as the FS-DEMO cables are compacted inside a stainless steel conduit.

Table 7.5: Full-size HTS conductors compared to DEMO requirements.

parameters	DEMO	FS-STC	FS-TSTC	FS-RACC	FS-CORC
cable space inside conduit (mm^2)	968.6	947.1	968.6	947.1	968.6
total copper stabilisation (mm^2)	554.9	598.8	592.7	598.8	414.8
copper former (mm^2)	-	448.3	159.1	448.3	145.5
copper superstrands (mm^2)	-	-	283.1	-	117.8
copper in *RE*BCO (mm^2)	-	150.5	150.5	150.5	151.5
entire helium void (mm^2)	272.2	272.2	272.8	272.2	272.2
entire stainless steel (mm^2)	1147.6	1146.2	1147.6	1146.2	1147.6
stainless steel in conduit (mm^2)	1056.2	1077.8	1056.3	1077.8	1056.2
stainless steel in *RE*BCO (mm^2)	91.4	68.4	68.4	68.4	68.9
stainless steel in former (mm^2)	-	-	22.9	-	22.5
insulation (mm^2)	300	300	300	300	300
operating current I_{op}(13.27 T, 5.0 K) (kA)	50	50.3	50.3	50.3	50.6
critical current I_c(13.27 T, 5.0 K) (kA)	71.4	71.8	71.8	71.8	72.3
safety factor (I_{op}/I_c)	0.7	0.7	0.7	0.7	0.7
number of strands (stacks / RACCs)	-	18	18	18	17
number of *RE*BCO tapes per strand	-	19	19	38	11, 12
*RE*BCO tape width (mm)	-	4	4	2	3, 4
strand transposition length (mm)	-	-	-	262	249
strand twist pitch (mm)	-	389	320	389	264.9
cable twist pitch (mm)	-	1002	1050	1002	765
cable strain (%)	-	< 0.6	< 0.6	< 0.6	-
*RE*BCO conductor length (km)	-	84,300	84,200	101,100	120,300

8 Summary, conclusion and outlook

The DEMOnstration reactor DEMO is planned to be the first prototype for a commercial fusion power plant. EUROfusion's offical schedule foresee first fusion electricity by 2050. According to the July 2012 design, the DEMO reactor requires 21% larger toroidal field coils, a 29% higher magnetic field and a 110% higher coil current than the actual ITER experiment. This engineering challenge has to be met with new technological developments. While ITER's plasma confining toroidal field coils are manufactured from low temperature superconductors, DEMO could benefit from 2G *RE*BCO high temperature superconductor technology. *RE*BCO coated conductors offer a higher engineering current density, lower magnetic field dependency, an increased tensile strength, high critical temperature limits and hence an increased current sharing temperature. As coated conductors come as flat tapes that cannot be wired as former low temperature Nb_3Sn superconductors, new cable designs and cabling methods are developed to manufacture full-size (FS) fusion conductor cables.

Main focus of this thesis is the investigation and evaluation of present high temperature superconductor cable concepts for their applicability in toroidal field coils of the DEMO power plant. Three different cable concepts and nine different cable samples were investigated, experimentally tested, evaluated and adapted to the DEMO input parameters. This includes two Roebel Assembled Coated Conductor (RACC) samples, six Stacked Tape Cable (STC) samples in twisted in non-twisted variant and one Conductor on Round Core (CORC) sample. All sample tests were executed in the FBI-T measurement facility where the critical current can be measured as function of the tensile force, the magnetic field and the sample temperature.

Before any sample was manufactured or prepared for in-field measurements, important data on soft-soldering of *RE*BCO superconductors from publications and experiments over the last decade was gathered and analysed. Soft soldering is the core process during manufacturing STCs and TSTCs and a promising approach to extend the single tape length from few hundred meters to many kilometres. It was shown that single tape joints are highly reproducible with specific joint resistances around $47.73 \pm 5.06\,\mathrm{n\Omega cm^2}$ for SnPb and $102.68 \pm 5.25\,\mathrm{n\Omega cm^2}$ for InSn. Both soft solders are qualified to be used for fusion applications as the constituents Pb, Sn and In satisfy the recycling limit of 100 years after nuclear activation, considering first wall dose simulations of the ITER experiment. During the soldering process the joint specific resistance can be improved significantly by reducing the solder layer thickness with a recommended transversal pressure of 5 MPa. Flux is a helpful tool to guarantee for a good connection between solder and the *RE*BCO tapes. However, only flux on base of colophony resin is recommended. Acid flux destroys the *RE*BCO conductors irreversibly. Longitudinal strain, as it is expected due to coil hoop stress or strand winding during cable assembly, does not impair face-to-face lap joints. Lap joints have a higher tensile strength than single *RE*BCO tapes. According to these soft-solder recommendations, it is possible to manufacture single tape extensions for 4 mm wide *RE*BCO tapes of less than 50 nΩ. The resulting heat input was calculated in chapter 7 for each full-size cable individually.

The main part of this thesis was the preparation and in-field investigation of the nine sub-size HTS cable samples. Special focus laid on the RACC, which was measured inside a magnetic background field of up to 12 T. Firstly, the Roebel strands of the RACC had to be mechanically stabilised against tape movement which lead to full cable degradation during former measurements. A proper solution to suppress tape movement was found by compacting the cable inside a pre-compression conduit. This implied an investigation on the maximal applicable transverse pressure. The found heterogeneous pressure distribution on the RACC plane damaged the strands at the tape crossovers irreversibly. By inserting additional copper strands inside the RACC, the pressure was distributed more homogeneously and the minimum pressure before degradation was extended from 90 to 120 MPa. In the following, two RACC samples were manufactured, compacted and tested successfully in magnetic background fields. The conduit proved to protect the RACC from Lorentz forces of 1.56 kN, which corresponds to a pressure of 17.33 MPa perpendicular to the cable plane. Beside the RACC three STCs, three TSTCs and one CORC were investigated for their field dependent current function at 4.2 K and their stability against Lorentz forces. Except for sample STC C, which was the only sample with a full-size former but degraded by 8%, all other samples withstood the experienced Lorentz forces and withstood a pressure acting perpendicular on the tape plane of approximately 20 MPa. With 10.7 kN the highest Lorentz force was by far experienced by sample TSTC F. This corresponds to a single tape pressure of 23 MPa perpendicular to the tape plane in the high-field zone. The only sample which could not be successfully tested was the CORC cable. The cable was fused during the 10 T measurement. Insufficient mechanical stabilisation lead to cable quench and the relatively bad heat exchange between the CORC strands made it impossible to avoid the double burnout. For the three most promising designs, one STC and two TSTCs, the critical current was measured at different magnetic fields and temperatures. While the first two samples met the theoretical expectations, the third sample suffered from a very inhomogeneous temperature distribution. For temperatures below 50 K, the very high thermal conductivity peak of the oxygen-free copper conduit made it impossible to adjust the stack temperature to a fix value. The deviation between measured and expected temperature implied a huge temperature gradient inside the stack. In an ultimate test one STC sample was investigated for the critical current curve depending on tensile strain in longitudinal direction at 4.2 K and 12 T background field. The acquired data was in conformity with the reference curve, taken from single tape measurements. The maximum peak strain before degradation was 0.6%. As copper and soft solder were the main constituents of the investigated sample B, the stress-strain diagram came close to annealed bronze, which is a Cu90Sn10 alloy. For fully equipped cable strands, with a higher stainless steel ratio, the stress-strain curve of a (T)STC strand will come closer to the single tape curve.

Based on the acquired cable data and the experience gathered with the nine sub-size cable samples, each of the four cable designs was adapted to the DEMO requirements, beginning with calculations on flux jump stability to prove wide *RE*BCO conductors a valid alternative to the existing Nb_3Sn Cable In Conduit Conductor (CICC) technology. Afterwards, each cable concept was up-scaled individually, taking the DEMO requirements as input parameters. The most promising full-size cable concepts are the FS-STC and the FS-TSTC. While the STC is a straight forward approach with a very low strand twist pitch, the TSTC comes with a higher strand twist for reduced AC losses. Both designs satisfy all DEMO requirements and the soldering of *RE*BCO tapes into a copper former is proven a very successful method to manufacture electromechanically very stable conductors. Although the required conductor length is far beyond actual single tape lengths, the heat input of the lap joints to extend the tape length is negligible. For both cable designs the required length of 4 mm wide *RE*BCO conductor is about 84,000 km. If future investigations and simulations prove that full tape transposition is required to successfully ramp the TF coils, the FS-RACC is the only option. If full transposition is not mandatory, the RACC should not be considered for fusion applications. Its filigree Roebel strands make the RACC very vulnerable to Lorentz forces and if not fully soldered into a copper former, the current sharing between the RACC and the copper is not yet sufficient. A cable quench might lead to a burnout of the RACC strands. Same accounts for the FS-CORC. Although it has a high resilience against hoop or winding strain, the layer winding of the tapes does not allow for an optimum heat exchange and current sharing between the tapes and the copper former. Both, heat and current have to pass the high ohmic stainless steel substrate of the *RE*BCO layers. Due the punched Roebel strands and the 45° winding of the CORC, the FS-RACC and the FS-CORC require 100,000 km and 120,000 km of 4 mm wide *RE*BCO conductor respectively.

Although this thesis closes with four feasible full-size cable propositions, many topics need to be addressed before a first full-size cable prototype could be manufactured. This includes modifications on the forced flow cooling channels, simulations of AC losses during magnet ramp up and an agreement over the final European DEMO design.

A Annex

A.1 Soldering contact database

In chapter 4 soft soldering contacts were evaluated and discussed. The large number on publications makes it difficult not to lose focus. Therefore, table **??** gives an overview of all gathered information. Since each publication follows a different approach, the table considers various process parameters. Gaps occur due to the fact, that not every publication does provide information on each parameter considered in the table. In case a publication did not mention a parameter explicitly, we tried to contact the authors. In case information could not be obtained, it was calculated on basis of the known parameters and is given in brackets.

Following abbreviations were used in the database.

A_c	contact area / mm²
l_c	contact length / mm
n	sample size
R_j	joint resistance at 77 K / Ω
r_{js}	specific joint resistance at 77 K / Ωmm^2
u	standard uncertainty
t	stabiliser thickness / mm
T	soldering temperature / °C
T_l	liquidus temperature / °C
A260	Alpha 260 HF
K135	Kester 135

Table A.1: Database on soft solders for high temperature superconductors

Solder	Composition	T_l	Solder type	Flux	R_j(77 K)	u	r_{js}(77 K)	SEM	n	l_c	A_c	Contact alignment
	m%	°C			nΩ	nΩ	nΩcm²	nΩcm²		mm	mm²	
InBi	In66Bi33	71			0.5		2.16		1	100	432	Lap, HTS-HTS
InBi	In66Bi33	71			1.6		3.2		1	50	200	Lap, HTS-HTS
InBi	In66Bi33	71			3		3.6		1	30	120	Lap, HTS-HTS
InBi	In66Bi33	71			4.3		3.6		1	20	83.8	Lap, HTS-HTS
InBi	In66Bi33	71			3.1		3.72		1	30	120	Lap, HTS-HTS
InBi	In66Bi33	71			9.4		3.76		1	10	40	Lap, HTS-HTS
InBi	In66Bi33	71			4.7		3.76		1	20	80	Lap, HTS-HTS
InBi	In66Bi33	71			2.4		3.84		1	40	160	Lap, HTS-HTS
InBi	In66Bi33	71			3.2		3.84		1	30	120	Lap, HTS-HTS
InBi	In66Bi33	71			2		4		1	50	200	Lap, HTS-HTS
InBi	In66Bi33	71			10		4		1	10	40	Lap, HTS-HTS
InBi	In66Bi33	71			3		5.18		1	40	172.8	Lap, HTS-HTS
InBi	In66Bi33	71			14		6.05		1	10	43.2	Lap, HTS-HTS
InBi	In66Bi33	71			14		6.05		1	10	43.2	Lap, HTS-HTS
InBi	In66Bi33	71			8		6.4		1	20	80	Lap, HTS-HTS
InBi	In66Bi33	71			8		6.91		1	20	86.4	Lap, HTS-HTS
InBi	In66Bi33	71			2		6.91		1	80	345.6	Lap, HTS-HTS
InBi	In66Bi33	71			16		6.91		1	10	43.2	Lap, HTS-HTS
InBi	In66Bi33	71			3		7.78		1	60	259.2	Lap, HTS-HTS
InBi	In66Bi33	71			10.4		8.32		1	20	80	Lap, HTS-HTS
InBi	In66Bi33	71			20		8.64		1	10	43.2	Lap, HTS-HTS
SnPb			Paste		4.52		9.04		1	50	200	Lap, HTS-HTS
InBi	In66Bi33	71			23		9.2		1	10	40	Lap, HTS-HTS
InBi	In66Bi33	71			23		9.94		1	10	43.2	Lap, HTS-HTS
InBi	In66Bi33	71			5.8		10.07		1	40	173.6	Lap, HTS-Substrate
InBi	In66Bi33	71			12.8		11.11		1	20	86.8	Lap, HTS-Substrate
InBi	In66Bi33	71			7.3		14.6		1	50	200	Bridge, HTS-HTS
InBi	In66Bi33	71			7.6		15.2		1	50	200	Bridge, HTS-HTS
SnPb			Paste		7.01		15.25		1	50	217.5	Lap, HTS-HTS
InBi	In66Bi33	71			9.6		15.36		1	40	160	Bridge, HTS-HTS
InBi	In66Bi33	71			9.7		15.52		1	40	160	Bridge, HTS-HTS
InBi	In66Bi33	71			21.3		17.04		1	20	80	Bridge, HTS-HTS
SnPb		183			2.3		17.4		1	63	762.3	Lap, HTS-HTS
InBi	In66Bi33	71			22.3		17.84		1	20	80	Bridge, HTS-HTS
InBi	In66Bi33	71			15.3		18.36		1	30	120	Bridge, HTS-HTS
InBi	In66Bi33	71			15.4		18.48		1	30	120	Bridge, HTS-HTS
InBi	In66Bi33	71			48.2		19.28		1	10	40	Bridge, HTS-HTS
InBi	In66Bi33	71			49.2		19.68		1	10	40	Bridge, HTS-HTS
SnPb	Sn63Pb37	183			16.1		21		1	30	130	Lap, HTS-HTS
InSn	In52Sn48	116			17.6		23		1	30	131	Lap, HTS-HTS
SnPb		180	Paste		11.95		23.9		1	50	200	Lap, HTS-HTS
SnPb		180	Paste		10		24		1	60	240	Lap, HTS-HTS

Heat source	T	Pretinning	Surface preparation	Pressure	Conductor	Producer	Stabiliser	t	Batch	Reference
	°C			MPa				μm		
Soldering iron				Applied		AMSC	Cu	40		[SDOB09]
Hot plate	120	Yes	Cleaned with ethyl alcohol	1	SCS4050	SP	Cu			[SD10]
Hot plate	120	Yes	Cleaned with ethyl alcohol	1	SCS4050	SP	Cu			[SD10]
Hot plate			Cleaned	1	YBCO	SP	Cu	20		[SDO10]
Hot plate	120	Yes	Cleaned with ethyl alcohol	1	SCS4050	SP	Cu			[SD10]
Hot plate	120	Yes	Cleaned with ethyl alcohol	1	SCS4050	SP	Cu			[SD10]
Hot plate	120	Yes	Cleaned with ethyl alcohol	1	SCS4050	SP	Cu			[SD10]
Hot plate	120	Yes	Cleaned with ethyl alcohol	1	SCS4050	SP	Cu			[SD10]
Hot plate	120	Yes	Cleaned with ethyl alcohol	1	SCS4050	SP	Cu			[SD10]
Hot plate	120	Yes	Cleaned with ethyl alcohol	1	SCS4050	SP	Cu			[SD10]
Hot plate	120	Yes	Cleaned with ethyl alcohol	1	SCS4050	SP	Cu			[SD10]
Soldering iron				Applied		AMSC	Cu	40		[SDOB09]
Hot plate				5		AMSC	Cu	42		[SDOB09]
Hot plate				10		AMSC	Cu	43		[SDOB09]
Hot plate	120	Yes	Cleaned with ethyl alcohol	1	SCS4050	SP	Cu			[SD10]
Soldering iron				Applied		AMSC	Cu	40		[SDOB09]
Soldering iron				Applied		AMSC	Cu	40		[SDOB09]
Hot plate				1		AMSC	Cu	41		[SDOB09]
Soldering iron				Applied		AMSC	Cu	40		[SDOB09]
Hot plate			Cleaned	1	SmBCO	KERI	Cu	20		[SDO10]
Hot plate				0.1		AMSC	Cu	40		[SDOB09]
Hot plate	190			Applied	344C	AMSC	x			[CKP+08]
Hot plate	120	Yes	Cleaned with ethyl alcohol	1	SCS4050	SP	Cu			[SD10]
Soldering iron				Applied		AMSC	Cu	40		[SDOB09]
Soldering iron				Applied		SP	Cu	20		[SDOB09]
Soldering iron				Applied		SP	Cu	20		[SDOB09]
Hot plate	120	Yes	Cleaned with ethyl alcohol	1	SCS4050	SP	Cu			[SD10]
Hot plate	120	Yes	Cleaned with ethyl alcohol	1	SCS4050	SP	Cu			[SD10]
Hot plate	190			Applied	344C	AMSC	Cu	50		[CKP+08]
Hot plate	120	Yes	Cleaned with ethyl alcohol	1	SCS4050	SP	Cu			[SD10]
Hot plate	120	Yes	Cleaned with ethyl alcohol	1	SCS4050	SP	Cu			[SD10]
Hot plate	120	Yes	Cleaned with ethyl alcohol	1	SCS4050	SP	Cu			[SD10]
Heater block		Yes	As manufactured	Applied	SCS12050	SP	Cu			[BOB+13]
Hot plate	120	Yes	Cleaned with ethyl alcohol	1	SCS4050	SP	Cu			[SD10]
Hot plate	120	Yes	Cleaned with ethyl alcohol	1	SCS4050	SP	Cu			[SD10]
Hot plate	120	Yes	Cleaned with ethyl alcohol	1	SCS4050	SP	Cu			[SD10]
Hot plate	120	Yes	Cleaned with ethyl alcohol	1	SCS4050	SP	Cu			[SD10]
Hot plate	120	Yes	Cleaned with ethyl alcohol	1	SCS4050	SP	Cu			[SD10]
	200			Applied		AMSC	x			[BOSF09]
	150			Applied		AMSC	x			[BOSF09]
Heater block	200			1.47	SCS4050	SP	Cu	40		[CPY+10]
Heater block	190			166.71	SCS4050	SP	Cu	20		[CJK+11]

Solder	Composition m%	T_l °C	Solder type	Flux	R_j(77 K) nΩ	u nΩ	$r_{j,s}$(77 K) nΩcm^2	SEM nΩcm^2	n	l_c mm	A_c mm^2	Contact alignment
SnAgCu	Sn96.5Ag3Cu0.5	217	Paste		8.35		26.72		1	80	(3.2)	Lap, HTS-HTS
In	In100	157		A260, K135	26	2.04	28.6	2.25	6	25	110	Lap, HTS-HTS
InBi	In66Bi33	71			13		28.6		1	50	220	Lap, HTS-HTS
SnPb		180	Paste		14.4		28.8		1	50	200	Lap, HTS-HTS
SnAg	Sn96.5Ag3.5	221			30	4.08	30.48	4.15	6	25.4	101.6	Lap, HTS-HTS
InBi	In66Bi33	71			23.4		30.89		1	30	132	Lap, HTS-HTS
SnAg	Sn94Ag4	221		Spirflux 330	13.77	1.77	32.47	1.38	4	2.03	244	Lap, HTS-HTS
SnPb		183			(80.4)		32.48		1	10.1	(40.4)	Lap, HTS-HTS
SnPb	Sn60Pb40	191	Ribbon 250 µm	Rosin flux	(100)		32.7		1	25	(100)	Lap, HTS-HTS
SnPb		183			(64.19)		33.38		1	13	(52)	Lap, HTS-HTS
SnPb		180	Paste		16.75		33.5		1	50	200	Lap, HTS-HTS
SnPb		180	Paste		7		33.6		1	120	480	Lap, HTS-HTS
SnPb		183			(72.5)		33.64		1	11.6	(46.4)	Lap, HTS-HTS
SnPb	Sn60Pb40	191	Ribbon 250 µm	Rosin flux	(100)		33.67		1	25	(100)	Lap, HTS-HTS
SnPb	Sn60Pb40	191	Ribbon 250 µm	Rosin flux	(100)		34.27		1	25	(100)	Lap, HTS-HTS
SnPb		183			(67.7)		34.39		1	12.7	(50.8)	Lap, HTS-HTS
SnPb	Sn60Pb40	191	Ribbon 250 µm	Rosin flux	(100)		34.94		1	25	(100)	Lap, HTS-HTS
InSn	In52Sn48	118		A260, K135	32	4.08	35.2	4.49	6	25	110	Lap, HTS-HTS
SnPb		180	Paste		18		36		1	50	200	Lap, HTS-HTS
SnPb		180	Paste		18.1		36.2		1	50	200	Lap, HTS-HTS
SnPb	Sn60Pb40	191	Ribbon 250 µm	Rosin flux	(100)		36.22		1	25	(100)	Lap, HTS-HTS
In	In100	157			23		(36.8)		1	40	(160)	Lap, HTS-HTS
SnPb	Sn60Pb40	191	Ribbon 250 µm	Rosin flux	(100)		36.89		1	25	(100)	Lap, HTS-HTS
SnPb		183			(83.91)		36.92		1	11	(44)	Lap, HTS-HTS
SnPb	Sn60Pb40	191			30		37		1	31	123.33	Lap, HTS-HTS
SnPb	Sn63Pb37	183	Ribbon		23.13		37		1	40	160	Lap, HTS-HTS
SnPb		183			(65.1)		37.5		1	14.4	(57.6)	Lap, HTS-HTS
SnPb	Sn60Pb40	191	Ribbon 250 µm	Rosin flux	(100)		38.16		1	25	(100)	Lap, HTS-HTS
Indalloy	In80Pb15Ag5	135	Round wire	CW8100	38.8	1.88	38.8	1.88	4	25	100	Lap, HTS-HTS
SnPb	Sn63Pb37	183	Ribbon		25		40		1	40	160	Lap, HTS-HTS
SnPb		183			(80)		40		1	12.5	(50)	Lap, HTS-HTS
SnPb		183			16.2		41.8		1	63	252	Lap, HTS-HTS
SnPb	Sn40Pb60	238			98.7		42.5		1		43.06	Bridge, HTS-HTS
SnPb		180	Paste		21.3		42.6		1	50	200	Lap, HTS-HTS
SnPb	Sn60Pb40	191	Ribbon 250 µm	Rosin flux	(100)		42.72		1	25	(100)	Lap, HTS-HTS
SnPb	Sn60Pb40	191			21		43		1	51	204.76	Lap, HTS-HTS
SnPb		183			(77)		43.12		1	14	(56)	Lap, HTS-HTS
SnPb		183			(83.6)		43.47		1	13	(52)	Lap, HTS-HTS
SnPb	Sn63Pb37	183	Ribbon		27.5		44		1	40	160	Lap, HTS-HTS
SnPb		180	Paste		22.35		44.7		1	50	200	Lap, HTS-HTS
SnPb	Sn60Pb40	191			53		45		1	21	84.91	Lap, HTS-HTS
Indalloy	In80Pb15Ag5	135	Round wire	CW8100	11.33	1.77	45.32	7.07	3	100	400	Lap, HTS-HTS

Heat source	T °C	Pretinning	Surface preparation	Pressure MPa	Conductor	Producer	Stabiliser	t μm	Batch	Reference
Soldering iron	250	No	Sand paper 15.3 μm particles	10	SCS4050	SP	Cu	20		[LZL+14]
Soldering iron	165	Yes	Cleaned with ethanol/methanol	Applied	344C	AMSC	Cu	80		[ZDHG11]
	160				344C	AMSC	Cu	50		[CPY+09]
	200			1.47	SCS4050	SP	Cu	40		[CPY+10]
Soldering iron	226-231	Yes	As manufactured	Applied	SCS4050	SP	Cu	20		[DZG+10]
	160				344C	AMSC	Cu	50		[CPY+09]
	230-240	Yes	Citric acid. purified water. ethanol	10Nm. M8 Screw	SCS12050	SP	Cu			[Bay13]
	200	Yes	nitric acid, 70 °C	9.7	SCS4050 AP	SP	Cu			[Cel13]
Hot plate	<250	No	Cleaned	Applied	SCS4050	SP	Cu	20		[TAB+12]
	210	Yes	nitric acid, 70 °C	7.54	SCS4050 AP	SP	Cu			[Cel13]
	200			1.47	SCS4050	SP	Cu	40		[CPY+10]
	190			166.71	SCS4050	SP	Cu	20		[CJK+11]
	210	Yes	nitric acid, 70 °C	8.45	SCS4050 AP	SP	Cu			[Cel13]
Hot plate	<250	No	Cleaned	Applied	SCS4050	SP	Cu	20		[TAB+12]
Hot plate	<250	No	Cleaned	Applied	SCS4050	SP	Cu	20		[TAB+12]
	200	Yes	nitric acid, 70 °C	7.72	SCS4050 AP	SP	Cu			[Cel13]
Hot plate	<250	No	Cleaned	Applied	SCS4050	SP	Cu	20		[TAB+12]
	150	Yes	Cleaned with ethanol/methanol	Applied	344C	AMSC	Cu	80		[ZDHG11]
	200			2.45	SCS4050	SP	Cu	40		[CPY+10]
	200			2.45	SCS4050	SP	Cu	40		[CPY+10]
Hot plate	<250	No	Cleaned	Applied	SCS4050	SP	Cu	20		[TAB+12]
				Applied	SCS4050	SP	Cu			[KKC+09]
Hot plate	<250	No	Cleaned	Applied	SCS4050	SP	Cu	20		[TAB+12]
	200	Yes	nitric acid, 70 °C	8.91	SCS4050 AP	SP	Cu			[Cel13]
	<200	Yes	Citric acid 5×10^{-3} M, pH 2.7, 170 °C	4.27	SCS4050	SP	Cu			[CF12]
	200	No		Applied	SCS4050	SP	Cu		SP7	[LHS+11]
	210	Yes	nitric acid, 70 °C	6.8	SCS4050 AP	SP	Cu			[Cel13]
Hot plate	<250	No	Cleaned	Applied	SCS4050	SP	Cu	20		[TAB+12]
Hot plate	155	No	Cleaning with acetone	0.09	SCS4050 AP	SP	Cu	5		[KBL+13]
	195			Applied	SCS4050	SP	Cu	50	SP26	[WMM+12]
	210	Yes	nitric acid, 70 °C	7.84	SCS4050 AP	SP	Cu			[Cel13]
		Yes	As manufactured	Applied	SCS4050	SP	Cu			[BOB+13]
	~200			Applied	344C	AMSC	Cu	123-147		[TYT+12]
	200			2.45	SCS4050	SP	Cu	40		[CPY+10]
Hot plate	<250	No	Cleaned	Applied	SCS4050	SP	Cu	20	B1	[TAB+12]
	<200		Citric acid 5×10^{-3} M, pH 2.7, 170 °C	2.59	SCS4050	SP	Cu			[CF12]
	210	Yes	nitric acid, 70 °C	7	SCS4050 AP	SP	Cu			[Cel13]
	210	Yes	nitric acid, 70 °C	7.54	SCS4050 AP	SP	Cu			[Cel13]
	200	No		Applied	SCS4050	SP	Cu		SP21	[LHS+11]
	200			2.45	SCS4050	SP	Cu	40		[CPY+10]
	<200		Citric acid 5×10^{-3} M, pH 2.7, 170 °C	6.3	SCS4050	SP	Cu			[CF12]
Hot plate	155		Cleaning with acetone	0.09	SCS4050 AP	SP	Cu	5		[KBL+13]

Solder	Composition m%	T_l °C	Solder type	Flux	R_j(77 K) nΩ	u nΩ	r_{js}(77 K) nΩcm^2	SEM nΩcm^2	n	l_c mm	A_c mm^2	Contact alignment
CerroLOW 136	Sn12Bi49Pb18In21	56			23.15		46.3		1	50	200	Lap, HTS-HTS
Indalloy	In80Pb15Ag5	158	Ribbon	CW8100	46.7	3.56	46.7	3.56	9	25	100	Lap, HTS-HTS
SnPb	Sn60Pb40	191			42		52		1	31	123.81	Lap, HTS-HTS
Indalloy	In80Pb15Ag5	135	Round wire	CW8100	26.33	0.88	52.66	1.77	3	50	200	Lap, HTS-HTS
Indalloy	In80Pb15Ag5	135	Round wire	CW8100	52.7	6.97	52.7	6.97	3	25	100	Lap, HTS-HTS
SnPb	Sn60Pb40	191	Ribbon 500 µm	Rosin flux	(100)		53.4		1	25	(100)	Lap, HTS-HTS
SnPb	Sn60Pb40	191			67		54		1	20	80.6	Lap, HTS-HTS
Indalloy	In80Pb15Ag5	135	Round wire	CW8100	54.29	4.89	54.29	4.89	7	25	100	Lap, HTS-HTS
SnPb		180	Paste		27.4		54.8		1	50	200	Lap, HTS-HTS
Indalloy	In80Pb15Ag5	135	Round wire	CW8100	18.33	2.19	54.99	6.57	3	75	300	Lap, HTS-HTS
Tix	Sn93In4.6Pb2.3	135	Round wire	CW8100	56.6	15.17	56.6	15.17	4	25	100	Lap, HTS-HTS
InBi	In66Bi33	71			71		56.8		1	20	80	Bridge, HTS-HTS
SnPb			Paste		28.95		57.9		1	50	200	Lap, HTS-HTS
SnPb	Sn60Pb40	191			47		58		1	31	123.4	Lap, HTS-HTS
Indalloy	In80Pb15Ag5	135	Round wire	CW8100	59.9	1.97	59.9	1.97	7	25	100	Lap, HTS-HTS
SnPb	Sn60Pb40	191			48		60		1	31	125	Lap, HTS-HTS
SnPb		183			(139)		60.05		1	10.8	(43.2)	Lap, HTS-HTS
SnAg	Sn96.5Ag3.5	221			60	12.25	60.96	12.44	6	25.4	101.6	Lap, HTS-HTS
SnPb	Sn63Pb37	183			38.13		61		1	40	160	Lap, HTS-HTS
SnPb	Sn63Pb37	183			38.75		62		1	40	160	Lap, HTS-HTS
SnPbIn				Acid based flux	(100)		62.51		1	25	(100)	Lap, HTS-HTS
SnPb	Sn60Pb40	191			41		64		1	39	156.1	Lap, HTS-HTS
SnPbIn				Acid based flux	(100)		65.06		1	25	(100)	Lap, HTS-HTS
Kester 44	Sn60Pb40	191	Round wire	CW8100	67	3.47	67	3.47	11	25	100	Lap, HTS-HTS
InBi					43		(68.8)		1	40	(160)	Lap, HTS-HTS
SnPb	Sn60Pb40	191			57		71		1	31	124.56	Lap, HTS-HTS
InSn	In52Sn48	118			70	24.5	71.12	24.89	6	25.4	101.6	Lap, HTS-HTS
InBi	In66Bi33	71			60		72		1	30	120	Bridge, HTS-HTS
SnPbIn				non-acid based flux	(100)		72.23		1	25	(100)	Lap, HTS-HTS
InSn		116			30.5		73		1	63	270.9	Lap, HTS-HTS
InSn	In52Sn48	118			60	8.17	73.15	9.95	6	25.4	121.92	Lap, HTS-HTS
SnPbBi	43Sn43P16Bi	163			60	12.25	73.15	14.93	6	25.4	121.92	Lap, HTS-HTS
In	In100	157			46		(73.6)		1	40	(160)	Lap, HTS-HTS
SnPb	Sn60Pb40	191			60		74		1	31	123.33	Lap, HTS-HTS
SnPb	Sn60Pb40	191			62		74		1	30	119.35	Lap, HTS-HTS
SnPb	Sn60Pb40	191			64		74		1	29	115.63	Lap, HTS-HTS
CerroLOW 136	Sn12Bi49Pb18In21	56			37.3		74.6		1	50	200	Lap, HTS-HTS
SnPbIn				Acid based flux	(100)		75.75		1	25	(100)	Lap, HTS-HTS
SnPb	Sn60Pb40	191			63		78		1	31	123.81	Lap, HTS-HTS
SnPb	Sn60Pb40	191			64		79		1	31	123.44	Lap, HTS-HTS
SnPb	Sn60Pb40	191			67		80		1	30	119.4	Lap, HTS-HTS
SnPb	Sn63Pb37	183	Ribbon		50		80		1	40	160	Lap, HTS-HTS

Heat source	*T* °C	Pretinning	Surface preparation	Pressure MPa	Conductor	Producer	Stabiliser	*t* μm	Batch	Reference
	75			2.45	SCS4050	SP	Cu	40		[CPY+10]
Hot plate	155	Yes	Cleaning with acetone	0.09	SCS4050 AP	SP	Cu	5		[KBL+13]
	<200		Citric acid 5×10^{-3} M, pH 2.7, 170 °C	4.27	SCS4050	SP	Cu			[CF12]
Hot plate	155		Cleaning with acetone	0.09	SCS4050 AP	SP	Cu	5		[KBL+13]
Hot plate	155	No	Cleaning with acetone	0.09	SCS4050 AP	SP	Cu	5		[KBL+13]
Hot plate	<250	No	Cleaned	Applied	SCS4050	SP	Cu	20	B2	[TAB+12]
	<200		Citric acid 5×10^{-3} M, pH 2.7, 170 °C	6.62	SCS4050	SP	Cu			[CF12]
Hot plate	155		Cleaning with acetone	0.09	SCS4050 AP	SP	Cu	5		[KBL+13]
	200			1.47	SCS4050	SP	Cu	40		[CPY+10]
Hot plate	155		Cleaning with acetone	0.09	SCS4050 AP	SP	Cu	5		[KBL+13]
Hot plate	155	Yes	Cleaning with acetone	0.09	SCS4050 AP	SP	Cu	5		[KBL+13]
	120	Yes	Cleaned with ethyl alcohol	1	SCS4050	SP	Cu			[SD10]
	190			Applied	344C	AMSC	x			[CKP+08]
	<200		Citric acid 5×10^{-3} M, pH 2.7, 170 °C	4.27	SCS4050	SP	Cu			[CF12]
Hot plate	155	No	Cleaning with acetone	0.09	SCS4050 AP	SP	Cu	5		[KBL+13]
	<200		Citric acid 5×10^{-3} M, pH 2.7, 170 °C	1.56	SCS4050	SP	Cu			[CF12]
	200	Yes	nitric acid, 70 °C	9.07	SCS4050 AP	SP	Cu			[Cel13]
	226-231	Yes	As manufactured	Applied	SCS4050	SP	Cu	20		[DZG+10]
	200	Yes		Applied	SCS4050	SP	Cu		SP21	[LHS+11]
	200	Yes		Applied	SCS4050	SP	Cu		SP7	[LHS+11]
Hot plate	<250		Cleaned	Applied	SCS4050	SP	Cu	20		[TAB+12]
	<200		Citric acid 5×10^{-3} M, pH 2.7, 170 °C	1.7	SCS4050	SP	Cu			[CF12]
Hot plate	<250		Cleaned	Applied	SCS4050	SP	Cu	20		[TAB+12]
Hot plate	185	Yes	Cleaning with acetone	0.09	SCS4050 AP	SP	Cu	5		[KBL+13]
				Applied	SCS4050	SP	Cu			[KKC+09]
Heater block	<200		Citric acid 5×10^{-3} M, pH 2.7, 170 °C	1.56	SCS4050	SP	Cu			[CF12]
Soldering iron	123-128	Yes	As manufactured	Applied	SCS4050	SP	Cu	20		[DZG+10]
Hot plate	120	Yes	Cleaned with ethyl alcohol	1	SCS4050	SP	Cu			[SD10]
Hot plate	<250		Cleaned	Applied	SCS4050	SP	Cu	20		[TAB+12]
Heater block		Yes	As manufactured	Applied	344C	AMSC	Cu			[BOB+13]
Soldering iron	123-128	Yes	As manufactured	Applied	348C	AMSC	Cu	50		[DZG+10]
Soldering iron	168-173	Yes	As manufactured	Applied	348C	AMSC	Cu	50		[DZG+10]
				Applied	344C	AMSC	Cu			[KKC+09]
Heater block	<200		Citric acid 5×10^{-3} M, pH 2.7, 170 °C	1.56	SCS4050	SP	Cu			[CF12]
Heater block	<200		Citric acid 5×10^{-3} M, pH 2.7, 170 °C	1.62	SCS4050	SP	Cu			[CF12]
Heater block	<200		Citric acid 5×10^{-3} M, pH 2.7, 170 °C	1.67	SCS4050	SP	Cu			[CF12]
Heater block	75			2.45	SCS4050	SP	Cu	40		[CPY+10]
Hot plate	<250		Cleaned	Applied	SCS4050	SP	Cu	20		[TAB+12]
Heater block	<200		Citric acid 5×10^{-3} M, pH 2.7, 170 °C	1.56	SCS4050	SP	Cu			[CF12]
Heater block	<200		Citric acid 5×10^{-3} M, pH 2.7, 170 °C	0.4	SCS4050	SP	Cu			[CF12]
Heater block	<200		Citric acid 5×10^{-3} M, pH 2.7, 170 °C	0.41	SCS4050	SP	Cu			[CF12]
	195			Applied	SCS4050	SP	Cu	38	SP7	[WMM+12]

Solder	Composition m%	T_l °C	Solder type	Flux	R_j(77 K) nΩ	u nΩ	$r_{j,s}$(77 K) nΩcm^2	SEM nΩcm^2	n	l_c mm	A_c mm^2	Contact alignment
SnPb					50		(80)		1	40	(160)	Lap, HTS-HTS
InBi	In66Bi33	71			206		82.4		1	10	40	Bridge, HTS-HTS
SnPb	Sn60Pb40	191			68		84		1	31	123.53	Lap, HTS-HTS
SnPbIn				Carboclic Acid	(100)		90.33		1	25	(100)	Lap, HTS-HTS
SnPb	Sn60Pb40	191			132		100		1	19	75.76	Lap, HTS-HTS
InBi					64		(102.4)		1	40	(160)	Lap, HTS-HTS
SnPb					65		(104)		1	40	(160)	Lap, HTS-HTS
SnAgCu	Sn96.5Ag3Cu0.5	220			5.3		106		1	20	2000	Lap, HTS-HTS
CerroLOW 136	Sn12Bi49Pb18In21	56			53.9		107.8		1	50	200	Lap, HTS-HTS
SnPb	Sn40Pb60	238			20		109.5		1		547.5	Bridge, HTS-HTS
CerroLOW 136	Sn12Bi49Pb18In21	56			55.25		110.5		1	50	200	Lap, HTS-HTS
SnPb					71		(113.6)		1	40	(160)	Lap, HTS-HTS
Indalloy	In80Pb15Ag5	135	Round wire	CW8100	116.5	5.71	116.5	5.71	4	25	100	Lap, HTS-HTS
Tix	Sn93In4.6Pb2.3	135	Round wire	CW8100	118.7	18.71	118.7	18.71	3	25	100	Lap, HTS-HTS
SnPbIn				Rosin flux	(100)		128.95		1	25	(100)	Lap, HTS-HTS
InBi	In66Bi33	71			31		133.92		1	100	432	Lap, HTS-Substrate
SnPb	Sn63Pb37	183	Ribbon		83.75		134		1	40	160	Lap, HTS-HTS
SnAgCu	Sn96.5Ag3Cu0.5	220			6.9		138		1	20	2000	Lap, HTS-HTS
CerroLOW 136	Sn12Bi49Pb18In21	56			70.55		141.1		1	50	200	Lap, HTS-HTS
CerroLOW 136	Sn12Bi49Pb18In21	56			70.95		141.9		1	50	200	Lap, HTS-HTS
CerroLOW 136	Sn12Bi49Pb18In21	56			73.1		146.2		1	50	200	Lap, HTS-HTS
InBi	In66Bi33	71			340		146.88		1	10	43.2	Lap, HTS-Substrate
InBi	In66Bi33	71			120		155.52		1	30	129.6	Lap, HTS-Substrate
InSn	In52Sn48	118		A260	75	2.04	165	4.49	6	50	220	Lap, HTS-HTS
InBi	In66Bi33	71			96		165.89		1	40	172.8	Lap, HTS-Substrate
SnPbBi	43Sn43P16Bi	163	Paste		150	12.25	167.64	13.69	6	25.4	111.76	Lap, HTS-HTS
SnPb	Sn63Pb37	183			106.88		171		1	40	160	Lap, HTS-HTS
InBi	In66Bi33	71			200		172.8		1	20	86.4	Lap, HTS-Substrate
InSn	In52Sn48	118		A260	55	2.04	181.5	6.74	6	75	330	Lap, HTS-HTS
InBi	In66Bi33	71			430		185.76		1	10	43.2	Lap, HTS-Substrate
InBi	In33Bi67	109			170	36.74	189.99	41.06	6	25.4	111.76	Lap, HTS-HTS
SnPb		183			47.7		190		1	32	387.2	Lap, HTS-HTS
SnPb	Sn63Pb37	183	Ribbon		120		192		1	40	160	Lap, HTS-HTS
SnPb		180	Paste		41		196.8		1	120	480	Bridge, HTS-HTS
InSn	In52Sn48	118		A260	180	8.17	198	8.98	6	25	110	Lap, HTS-HTS
SnPbCu	Sn60Pb38Cu2	190					200		1		25.8	Lap, HTS-Cu
InSn	In52Sn48	118			180	8.17	201.17	9.12	6	25.4	111.76	Lap, HTS-HTS
Indalloy	In80Pb15Ag5	135	Round wire	CW8100	205.3	9.28	205.3	9.28	5	25	100	Lap, HTS-HTS
SnPb	Sn63Pb37	183	Ribbon		128.75		206		1	40	160	Lap, HTS-HTS
SnPb		180	Paste		44		211.2		1	120	480	Bridge, HTS-HTS
In	In100	157			190	24.5	212.34	27.38	6	25.4	111.76	Lap, HTS-HTS
SnPb	Sn60Pb40	191		Rosin flux	(100)		215.79		1	25	(100)	Lap, HTS-HTS

Heat source	T	Pretinning	Surface preparation	Pressure	Conductor	Producer	Stabiliser	t	Batch	Reference
	°C			MPa				μm		
				Applied	SCS4050	SP	Cu			[KKC$^+$09]
Hot plate	120	Yes	Cleaned with ethyl alcohol	1	SCS4050	SP	Cu			[SD10]
Heater block	<200		Citric acid 5×10^{-3} M, pH 2.7, 170 °C	4.27	SCS4050	SP	Cu			[CF12]
Hot plate	<250		Cleaned	Applied	SCS4050	SP	Cu	20		[TAB$^+$12]
Heater block	<200		Citric acid 5×10^{-3} M, pH 2.7, 170 °C	1.74	SCS4050	SP	Cu			[CF12]
				Applied	344C	AMSC	Cu			[KKC$^+$09]
				Applied	344S	AMSC	Steel			[KKC$^+$09]
Heater block	250			Applied	YBCO		Ag	25		[SNS$^+$07]
Heater block	75			1.47	SCS4050	SP	Cu	40		[CPY$^+$10]
Hot plate	~200			Applied	344C	AMSC	Cu	123-147		[TYT$^+$12]
Heater block	75			2.45	SCS4050	SP	Cu	40		[CPY$^+$10]
				Applied	344C	AMSC	Cu			[KKC$^+$09]
Hot plate	155	No	Cleaning with acetone	0.09	SCS4050 AP	SP	Cu	5		[KBL$^+$13]
Hot plate	185	Yes	Cleaning with acetone	0.09	SCS4050 AP	SP	Cu	5		[KBL$^+$13]
Hot plate	<250		Cleaned	Applied	SCS4050	SP	Cu	20		[TAB$^+$12]
Soldering iron				Applied		AMSC	Cu	40		[SDOB09]
Heater block	200	No		Applied	SCS4050	SP	Cu		SP19	[LHS$^+$11]
Heater block	250			Applied	YBCO		Ag	25		[SNS$^+$07]
Heater block	75			1.47	SCS4050	SP	Cu	40		[CPY$^+$10]
Heater block	75			2.45	SCS4050	SP	Cu	40		[CPY$^+$10]
Heater block	75			1.47	SCS4050	SP	Cu	40		[CPY$^+$10]
Soldering iron				Applied		AMSC	Cu	40		[SDOB09]
Soldering iron				Applied		AMSC	Cu	40		[SDOB09]
Soldering iron	160	Yes	Cleaned with ethanol/methanol	Applied	344B	AMSC	Brass	150		[ZDHG11]
Soldering iron				Applied		AMSC	Cu	40		[SDOB09]
Heater block	168-173	Yes	As manufactured	Applied	344B	AMSC	Brass	110		[DZG$^+$10]
Heater block	200	Yes		Applied	SCS4050	SP	Cu		SP19	[LHS$^+$11]
Soldering iron				Applied		AMSC	Cu	40		[SDOB09]
Soldering iron	160	Yes	Cleaned with ethanol/methanol	Applied	344B	AMSC	Brass	150		[ZDHG11]
Soldering iron				Applied		AMSC	Cu	40		[SDOB09]
Soldering iron	114-119	Yes	As manufactured	Applied	344C	AMSC	Cu	50		[DZG$^+$10]
Heater block		Yes	As manufactured	Applied	SF12100	SP	x			[BOB$^+$13]
Heater block	200	No		Applied	SCS4050	SP	Cu		SP20	[LHS$^+$11]
Heater block	190			166.71	SCS4050	SP	Cu	20		[CJK$^+$11]
Soldering iron	150	Yes	Cleaned with ethanol/methanol	Applied	344B	AMSC	Brass	150		[ZDHG11]
	250-270						Cu			[FDRW$^+$10]
Soldering iron	123-128	Yes	As manufactured	Applied	344B	AMSC	Brass	110		[DZG$^+$10]
Hot plate	155	No	Cleaning with acetone	0.09	SCS4050 AP	SP	Cu	5		[KBL$^+$13]
Heater block	200	No		Applied	SCS4050	SP	Cu		SP20	[LHS$^+$11]
Heater block	190			166.71	SCS4050	SP	Cu	20		[CJK$^+$11]
Soldering iron	162-167	Yes	As manufactured	Applied	344B	AMSC	Brass	110		[DZG$^+$10]
Hot plate	<250	No	Cleaned	Applied		SP	x			[TAB$^+$12]

Solder	Composition m%	T_l °C	Solder type	Flux	R_j(77 K) nΩ	u nΩ	r_{js}(77 K) nΩcm^2	SEM nΩcm^2	n	l_c mm	A_c mm^2	Contact alignment
In	In100	157		A260, K135	200	24.5	220	26.94	6	25	110	Lap, HTS-HTS
InBi	In66Bi33	71			545		235.44		1	10	43.2	Lap, Sub-Sub
SnPb	Sn60Pb40	191		Rosin flux	(100)		237.84		1	25	(100)	Lap, HTS-HTS
Indalloy	In80Pb15Ag5	135	Round wire	CW8100	238.4	11.56	238.4	11.56	3	25	100	Lap, HTS-HTS
Indalloy	In80Pb15Ag5	135	Round wire	CW8100	241.8	3.77	241.8	3.77	3	25	100	Lap, HTS-HTS
Kester 44	Sn60Pb40	191	Round wire	CW8100	254.9	28.41	254.9	28.41	6	25	100	Lap, HTS-HTS
SnPb		183			(419.93)		257		1	15.3	(61.2)	Lap, HTS-HTS
CerroLOW 136	Sn12Bi49Pb18In21	56			128.7		257.4		1	50	200	Lap, HTS-HTS
SnPb		183			(563.03)		268		1	11.9	(47.6)	Lap, HTS-HTS
SnPb		183			(586.96)		270		1	11.5	(46)	Lap, HTS-HTS
SnPb		183			(586.96)		270		1	11.5	(46)	Lap, HTS-HTS
SnPb		183			(581.90)		270		1	11.6	(46.4)	Lap, HTS-HTS
SnPb		180	Paste		115		276		1	60	240	Bridge, HTS-HTS
Indalloy	In80Pb15Ag5	135	Round wire	CW8100	285	9.54	285	9.54	3	25	100	Lap, HTS-HTS
SnPb		183			(609.24)		290		1	11.9	(47.6)	Lap, HTS-HTS
SnPb		183			(725.25)		293		1	10.1	(40.4)	Lap, HTS-HTS
InBi	In66Bi33	71			138		300.15		1	50	217.5	Lap, HTS-HTS
SnPb		180	Paste		129		309.6		1	60	240	Bridge, HTS-HTS
SnPb		183			(586.47)		312		1	13.3	(53.2)	Lap, HTS-HTS
SnPb		183			55.4		335		1	49	592.9	Lap, HTS-HTS
SnPb		180	Paste		70		336		1	120	480	Bridge, HTS-HTS
SnPb		183			38.5		338		1	72	871.2	Lap, HTS-HTS
SnPb		180	Paste		145		348		1	60	240	Bridge, HTS-HTS
Indalloy	In80Pb15Ag5	135	Round wire	CW8100	370.4	36.67	370.4	36.67	4	25	100	Lap, HTS-HTS
In	In100	157			(842,98)		408		1	12.1	(48.4)	Lap, HTS-HTS
Indalloy	In80Pb15Ag5	135	Round wire	CW8100	499	69.25	499	69.25	5	25	100	Lap, HTS-HTS
Indalloy	In80Pb15Ag5	135	Round wire	CW8100	515.9	34.47	515.9	34.47	14	25	100	Lap, HTS-HTS
SnAg	Sn96.5Ag3.5	221			520	44.91	528.32	45.63	6	25.4	101.6	Lap, HTS-Substrate
InSn	In52Sn48	116			87		574		1	150	661	Lap, HTS-HTS
InSn		116			87		574		1	150	645	Lap, HTS-HTS
Indalloy	In80Pb15Ag5	135	Round wire	CW8100	145	8.66	580	34.64	3	100	400	Lap, HTS-HTS
Indalloy	In80Pb15Ag5	135	Round wire	CW8100	586.4	39.29	586.4	39.29	4	25	100	Lap, HTS-HTS
InSn	In52Sn48	116			191		588		1	70	308	Lap, HTS-HTS
InSn		116			191		588		1	70	301	Lap, HTS-HTS
SnPb		183			(1275.86)		592		1	11.6	(46,4)	Lap, HTS-HTS
SnPb		183			50.4		614		1	101	1222.1	Lap, HTS-HTS
InSn	In52Sn48	116			73		642		1	200	880	Lap, HTS-HTS
InSn		116			73		642		1	199	855.7	Lap, HTS-HTS
SnPb	Sn63Pb37	183	Bath		16.19		667.03		1	1030	(4120)	Bridge, HTS-HTS
SnPb		183			37.1		671		1	150	1815	Lap, HTS-HTS
InBi	In33Bi67	108		A260, K135	620	232.70	682	255.97	6	25	110	Lap, HTS-HTS
SnPb	Sn63Pb37	183	Bath		77.73		684.02		1	220	(880)	Bridge, HTS-HTS

Heat source	T °C	Pretinning	Surface preparation	Pressure MPa	Conductor	Producer	Stabiliser	t μm	Batch	Reference
Soldering iron	165	Yes	Cleaned with ethanol/methanol	Applied	344B	AMSC	Brass	150		[ZDHG11]
Soldering iron				Applied		AMSC	Cu	40		[SDOB09]
Hot plate	<250	No	Cleaned	Applied		SP	x			[TAB+12]
Hot plate	155	No	Cleaning with acetone	0.09	SCS4050 AP	SP	Cu	5		[KBL+13]
Hot plate	155	No	Cleaning with acetone	0.09	SCS4050 AP	SP	Cu	5		[KBL+13]
Hot plate	185	Yes	Cleaning with acetone	0.09	SCS4050 AP	SP	Cu	5		[KBL+13]
Heater block	210	Yes	nitric acid, 70 °C	6.4	GdBCO	SuNAM	Cu			[Cel13]
Heater block	75			1.47	SCS4050	SP	Cu	40		[CPY+10]
Heater block	200	Yes	nitric acid, 70 °C	8.24	GdBCO	SuNAM	Cu			[Cel13]
Heater block	210	Yes	nitric acid, 70 °C	8.52	GdBCO	SuNAM	Cu			[Cel13]
Heater block	210	Yes	nitric acid, 70 °C	8.52	GdBCO	SuNAM	Cu			[Cel13]
Heater block	210	Yes	nitric acid, 70 °C	8.45	GdBCO	SuNAM	Cu			[Cel13]
Heater block	190			166.71	SCS4050	SP	Cu	20		[CJK+11]
Hot plate	155	No	Cleaning with acetone	0.09	SCS4050 AP	SP	Cu	5		[KBL+13]
Heater block	200	Yes	nitric acid, 70 °C	8.23	GdBCO	SuNAM	Cu			[Cel13]
Heater block	200	Yes	nitric acid, 70 °C	9.7	GdBCO	SuNAM	Cu			[Cel13]
Hot plate	80			Applied	344C	AMSC	Cu	50		[PAK+07]
Heater block	190			166.71	SCS4050	SP	Cu	20		[CJK+11]
Heater block	200	Yes	nitric acid, 70 °C	7.37	GdBCO	SuNAM	Cu			[Cel13]
Heater block		Yes	As manufactured	Applied	SF12100	SP	x			[BOB+13]
Heater block	190			166.71	SCS6050	SP	Cu	20		[CJK+11]
Heater block		Yes	As manufactured	Applied	SF12100	SP	x			[BOB+13]
Heater block	190			166.71	SCS6050	SP	Cu	20		[CJK+11]
Hot plate	155	No	Cleaning with acetone	0.09	SCS4050 AP	SP	Cu	5		[KBL+13]
Heater block		Yes	nitric acid, 70 °C	6.48	Amperium	AMSC	Cu			[Cel13]
Hot plate	155		Cleaning with acetone	0.09	SCS4050 AP	SP	Cu	5		[KBL+13]
Hot plate	155	No	Cleaning with acetone	0.09	SCS4050 AP	SP	Cu	5		[KBL+13]
Soldering iron	226-231	Yes	As manufactured	Applied	SCS4050	SP	Cu	20		[DZG+10]
	200			Applied		AMSC	Steel			[BOSF09]
Heater block		Yes	As manufactured	Applied	344S	AMSC	x			[BOB+13]
Hot plate	155		Cleaning with acetone	0.09	SCS4050 AP	SP	Cu	5		[KBL+13]
Hot plate	155	No	Cleaning with acetone	0.09	SCS4050 AP	SP	Cu	5		[KBL+13]
	200			Applied		AMSC	Steel			[BOSF09]
Heater block		Yes	As manufactured	Applied	344S	AMSC	x			[BOB+13]
Heater block		Yes	nitric acid, 70 °C	8.45		SuperOx	Cu			[Cel13]
Heater block		Yes	As manufactured	Applied	SF12100	SP	x			[BOB+13]
	200			Applied		AMSC	Steel			[BOSF09]
Heater block		Yes	As manufactured	Applied	344S	AMSC	x			[BOB+13]
Hot bath						SuNAM	Cu			[MZC+15]
Heater block		Yes	As manufactured	Applied	SF12100	SP	x			[BOB+13]
Soldering iron	150	Yes	Cleaned with ethanol/methanol	Applied	344C	AMSC	Cu	80		[ZDHG11]
Hot bath						SuNAM	Cu			[MZC+15]

Solder	Composition m%	T_l °C	Solder type	Flux	R_j(77 K) nΩ	u nΩ	r_{js}(77 K) nΩcm^2	SEM nΩcm^2	n	l_c mm	A_c mm^2	Contact alignment
InBi	In33Bi67	109			620	236.78	692.91	264.63	6	25.4	111.76	Lap, HTS-HTS
InBi	In33Bi67	108		A260, K135	630	204.12	693	224.54	6	25	110	Lap, HTS-HTS
InSn	In52Sn48	116			158		695		1	100	440	Lap, HTS-HTS
InSn		116			158		695		1	100	430	Lap, HTS-HTS
SnPb	Sn63Pb37	183	Bath		2.91		698.4		1	6000	(24000)	Bridge, HTS-HTS
SnPb	Sn63Pb37	183	Bath		55.01		704.13		1	320	(1280)	Bridge, HTS-HTS
SnPb	Sn63Pb37	183	Bath		8.81		704.8		1	2000	(8000)	Bridge, HTS-HTS
SnPb	Sn63Pb37	183	Bath		2.25		720		1	8000	(32000)	Bridge, HTS-HTS
SnPb	Sn63Pb37	183	Bath		150.3		721.44		1	120	(480)	Bridge, HTS-HTS
SnPb	Sn63Pb37	183	Bath		4.55		728		1	4000	(16000)	Bridge, HTS-HTS
InSn	In52Sn48	118			720	24.5	731.52	24.89	6	25.4	101.6	Lap, HTS-Substrate
SnPb	Sn63Pb37	183	Bath		6.13		735.6		1	3000	(12000)	Bridge, HTS-HTS
SnPb	Sn63Pb37	183	Bath		36.49		744.4		1	510	(2040)	Bridge, HTS-HTS
SnPb		183			(1686.98)		803		1	11.9	(47.6)	Lap, HTS-HTS
Kester 44	Sn60Pb40	191	Round wire	CW8100	820.9	139.82	820.9	139.82	3	25	100	Lap, HTS-HTS
SnPb	Sn63Pb37	183			516.25		826		1	40	160	Lap, HTS-HTS
InSn	In52Sn48	116			667		880		1	30	132	Lap, HTS-HTS
InSn		116			667		880		1	30	129	Lap, HTS-HTS
SnPb	Sn63Pb37	183	Ribbon		593.75		950		1	40	160	Lap, HTS-HTS
SnPb	Sn63Pb37	183			596.88		955		1	40	160	Lap, HTS-HTS
SnPb	Sn63Pb37	183			608.75		974		1	40	160	Lap, HTS-HTS
SnPb	Sn63Pb37	183	Ribbon		631.25		1010		1	40	160	Lap, HTS-HTS
SnPb	Sn63Pb37	183	Ribbon		650		1040		1	40	160	Lap, HTS-HTS
SnPb	Sn63Pb37	183	Ribbon		690		1104		1	40	160	Lap, HTS-HTS
SnPb	Sn63Pb37	183	Ribbon		723.13		1157		1	40	160	Lap, HTS-HTS
In	In100	157			(2759.34)		1181		1	10.7	(42.8)	Lap, HTS-HTS
SnPb	Sn63Pb37	183			756.25		1210		1	40	160	Lap, HTS-HTS
SnPb	Sn63Pb37	183			763.75		1222		1	40	160	Lap, HTS-HTS
SnPbBi	43Sn43P16Bi	163			1010	126.56	1231.39	154.30	6	25.4	121.92	Lap, HTS-Substrate
InSn	In52Sn48	118			1040	114.31	1267.97	139.37	6	25.4	121.92	Lap, HTS-Substrate
SnPb	Sn63Pb37	183			800		1280		1	40	160	Lap, HTS-HTS
SnPb	Sn63Pb37	183			1091.25		1746		1	40	160	Lap, HTS-HTS
SnPb		183			(5909.84)		2884		1	12.2	(48.8)	Lap, HTS-HTS
In	In100	157			(12608.33)		3026		1	12	(48)	Lap, HTS-HTS
SnPbBi	43Sn43P16Bi	163	Paste		2720	118.39	3039.87	132.31	6	25.4	111.76	Lap, HTS-Substrate
In	In100	157			(15415.81)		6043		1	9.8	(39.2)	Lap, HTS-HTS
InSn	In52Sn48	118		A260, K135	5700	612.37	6270	673.61	6	25	110	Lap, HTS-HTS
In	In100	157		A260, K135	6200	775.67	6820	853.24	6	25	110	Lap, HTS-HTS
InBi	In66Bi33	71			3240		7047		1	50	217.5	Lap, Sub-Sub

Heat source	T °C	Pretinning	Surface preparation	Pressure MPa	Conductor	Producer	Stabiliser	t μm	Batch	Reference
Soldering iron	114-119	Yes	As manufactured	Applied	344B	AMSC	Brass	110		[DZG+10]
Soldering iron	140	Yes	Cleaned with ethanol/methanol	Applied	344B	AMSC	Brass	150		[ZDHG11]
	200			Applied		AMSC	Steel			[BOSF09]
Heater block		Yes	As manufactured	Applied	344S	AMSC	x			[BOB+13]
Hot bath						SuNAM	Cu			[MZC+15]
Hot bath						SuNAM	Cu			[MZC+15]
Hot bath						SuNAM	Cu			[MZC+15]
Hot bath						SuNAM	Cu			[MZC+15]
Hot bath						SuNAM	Cu			[MZC+15]
Hot bath						SuNAM	Cu			[MZC+15]
Soldering iron	123-128	Yes	As manufactured	Applied	SCS4050	SP	Cu	20		[DZG+10]
Hot bath						SuNAM	Cu			[MZC+15]
Hot bath						SuNAM	Cu			[MZC+15]
Heater block		Yes	nitric acid, 70 °C	8.24		SuperOx	Cu			[Cel13]
Hot plate	185	Yes	Cleaning with acetone	0.09	SCS4050 AP	SP	Cu	5		[KBL+13]
Heater block	200	Yes		Applied	SCS4050	SP	Cu		SP17	[LHS+11]
	200			Applied		AMSC	Steel			[BOSF09]
Heater block		Yes	As manufactured	Applied	344S	AMSC	x			[BOB+13]
Heater block	200	No		Applied	SCS4050	SP	Cu		SP17	[LHS+11]
Heater block	205	Yes		Applied	SCS4050	SP	Cu		SP17	[LHS+11]
Heater block	200	Yes		Applied	SCS4050	SP	Cu		SP17	[LHS+11]
Heater block	205	No		Applied	SCS4050	SP	Cu		SP17	[LHS+11]
Heater block	200	No		Applied	SCS4050	SP	Cu		SP17	[LHS+11]
Heater block	220	No		Applied	SCS4050	SP	Cu		SP17	[LHS+11]
Heater block	250	No		Applied	SCS4050	SP	Cu		SP17	[LHS+11]
Heater block	175	Yes	nitric acid, 70 °C	7.33	Amperium	AMSC	Cu			[Cel13]
Heater block	205	Yes		Applied	SCS4050	SP	Cu		SP17	[LHS+11]
Heater block	220	Yes		Applied	SCS4050	SP	Cu		SP17	[LHS+11]
Soldering iron	168-173	Yes	As manufactured	Applied	348C	AMSC	Cu	50		[DZG+10]
Soldering iron	123-128	Yes	As manufactured	Applied	348C	AMSC	Cu	50		[DZG+10]
Heater block	205	Yes		Applied	SCS4050	SP	Cu		SP17	[LHS+11]
Heater block	250	Yes		Applied	SCS4050	SP	Cu		SP17	[LHS+11]
Heater block	210	Yes	nitric acid, 70 °C	8.03	GdBCO	SuNAM	Cu			[Cel13]
Heater block	175	Yes	nitric acid, 70 °C	6.53	Amperium	AMSC	Cu			[Cel13]
Heater block	168-173	Yes	As manufactured	Applied	344B	AMSC	Brass	110		[DZG+10]
Heater block		Yes	nitric acid, 70 °C	8	Amperium	AMSC	Cu			[Cel13]
Soldering iron	165	Yes	Cleaned with ethanol/methanol	Applied	344S	AMSC	Steel	170		[ZDHG11]
Soldering iron	170	Yes	Cleaned with ethanol/methanol	Applied	344S	AMSC	Steel	170		[ZDHG11]
Hot plate	80			Applied	344C	AMSC	Cu	50		[PAK+07]

A.2 FBI-T sample specifications

Samples for the FBI-T measurement have to satisfy dimensional requirements. One of the following variations should be considered during the sample design in order to fit the dimensional boundary conditions. Following aspects account for any of the three variations:

- The sample terminations will be clamped by screws into the FBI-T terminals.
- The current enters the terminals only from one side.
- One pair of voltage taps is required to detect sample quenches.
- Up to four pairs of voltage taps can be used to determine the critical current I_c. One pair to cover the magnet high field area is recommended.
- In order to measure the sample temperature most precisely, it would be helpful to have one Cernox^{TM} SD temperature sensor inside the sample.

Variation 1: Both very long copper terminations are used to enter the current into the sample and to grip the sample to apply tensile load.

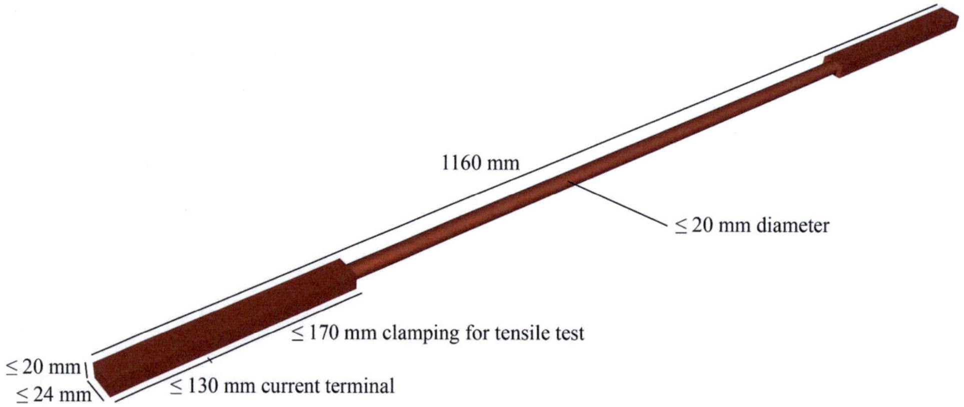

Figure A.1: Variation 1

Variation 2: The copper terminations are reduced increased in length and a stainless steel cuboid is attached to the sample to take the grip for the tensile test.

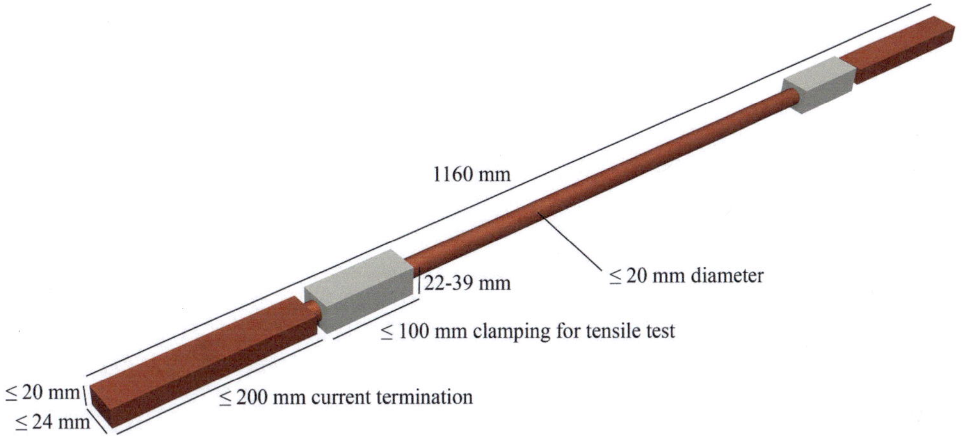

Figure A.2: Variation 2

A.3 Flux-jumping in a superconductor

Fluxjumping in a superconductor slab

When a minimum of heat ΔQ_{ext} is supplied to a superconductor, an external magnetic field will penetrate further into the superconductor. Main reason for this behaviour is the dependency of the critical current density j_c and the temperature T. In the following calculation we estimate a linear dependency where j_{c0} is the critical current at the lowest known temperature T_0:

$$j_c = \frac{j_{c0}(T_c - T)}{T_c - T_0} \tag{A.1}$$

By adding heat ΔQ_{ext} locally, the critical current density drops. Hitherto, superconducting screening currents become resistive currents, dissipating additional heat ΔQ_{int}. Since the screening currents are reduced, the magnetic field penetrates further into the superconductor. Figure A.3 demonstrates the magnetic field before and after the flux jump.

In the following we will calculate the generated heat going along with this flux jump on basis of the assumption that an infinitesimal width δx encapsulates an infinitesimal magnetic flux change $\delta\phi(x)$. For the internal heat ΔQ accounts over the cable width x:

$$\Delta Q_{int} = \int_a^b \delta q(x)\,\mathrm{d}x \tag{A.2}$$

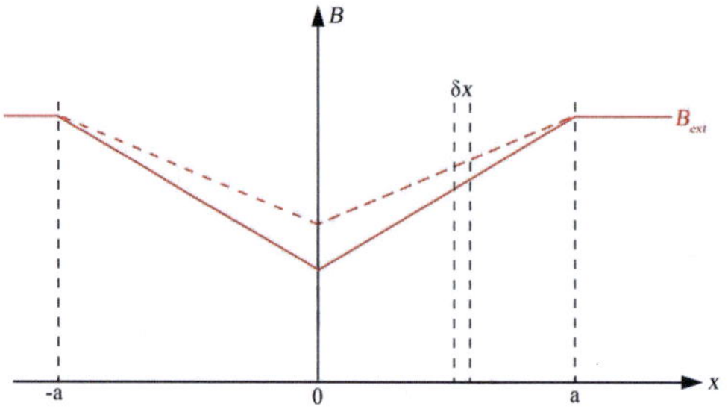

Figure A.3: External magnetic field penetrating a superconductor before (continuous line) and after the flux jump (dashed line).

Since the electric energy is dissipated in heat, the heat change $\delta q(x)$ can be calculated by integrating over the resistive power under the assumption, that within δx , the critical current density j_c is constant:

$$\delta q(x) = \int I(x)E(x)\,dt = j_c\,\delta x\,\delta\phi(x) \tag{A.3}$$

with

$$E(x) = \frac{\delta\phi(x)}{dt},\ I(x) = j_c\,\delta x.$$

To calculate the magnetic flux to solve equation A.3 we integrate the flux change over the x-direction with

$$\delta\phi(x) = \int_0^x \Delta B(x)\,dx = \int_0^x \mu_0\,\Delta j_c\,(a-x)dx = \mu_0\,\Delta j_c\,(ax - \frac{x^2}{2}) \tag{A.4}$$

having taken into account Ampère's circuital law and the fact, that the electric field along the conductor width is zero (E=0)

$$\boldsymbol{\nabla} \times \boldsymbol{B} = \mu_0 \boldsymbol{j} + \mu_0\varepsilon_0 \frac{\partial \boldsymbol{E}}{dt} \rightarrow \frac{\partial B_y}{\partial x} = \mu_0 j_c.$$

Combining formula A.2 and formula A.4 results in

$$\Delta Q_{int} = \frac{1}{a}\int_0^a \mu_0 j_c\,\Delta j_c(ax - \frac{x^2}{2})\,dx = \mu_0 j_c\,\Delta j_c \frac{a^2}{3}. \tag{A.5}$$

With our assumption from A.1, the rise in critical current Δj_c is

$$\Delta j_c = -j_c \frac{\Delta T}{T_c - T_0}$$

and the internal increase in heat becomes

$$\Delta Q_{int} = -\frac{\mu_0 j_c^2 a^2}{3(T_c - T_0)} \Delta T.$$

Taking into account the external heat ΔQ_{ext}, the effective specific heat per unit mass γC is

$$\gamma C_{eff} = \frac{\Delta Q_{ext} + \Delta Q_{int}}{\Delta T} = \frac{\Delta Q_{ext}}{\Delta T} - \frac{\mu_0 j_c^2 a^2}{3(T_c - T_0)} = \gamma C - \frac{\mu_0 j_c^2 a^2}{3(T_c - T_0)}$$

where γ is the density and C is the specific heat. It is obvious, that the energy of the screening currents (last term) reduces the specific heat per unit mass. Worst case scenario would be when the effective specific heat per mass γC_{eff} becomes zero. Any slight disturbance would lead to a tremendous rise in temperature and therefore to a spontaneous flux jump.

The minimum condition to prevent this incident is

$$0 = \gamma C - \frac{\mu_0 j_c^2 a^2}{3(T_c - T_0)}$$

with the defined stability parameter β

$$\beta = \frac{\mu_0 j_c^2 a^2}{\gamma C(T_c - T_0)} \overset{!}{\leq} 3.$$

For β values below 3, spontaneous flux jumps occur. The whole composition is adiabatic due to the fact, that there is no heat exchange with the environment.

Thermal and magnetic diffusion

The diffusion equation for a constant diffusion coefficient is

$$\frac{\partial n(x,t)}{\partial t} = D \Delta n(x,t).$$

In case of our superconductor the equations are

$$\frac{\partial T(x,t)}{\partial t} = D_T \Delta T(x,t),$$

$$\frac{\partial B(x,t)}{\partial t} = D_m \Delta B(x,t),$$

$$\frac{\partial j(x,t)}{\partial t} = D_m \Delta j(x,t),$$

where $D_m = \rho/\mu_0$ is the magnetic and $D_T = k/\gamma C$ is the thermal diffusivity. These differential equations have been solved for a variation of boundary conditions. The solution for linear heat flow in a solid within the boundary of two planes is described by Carslaw and Jaeger [QUELLE]. The dominant proportionality of the solution is

$$T, B, j \sim e^{\frac{-D\pi^2 t}{4a^2}}$$

where the characteristic time τ_j for the diffusion process is

$$\tau_j = \frac{4a^2}{\pi^2 D_m}$$

Flux-jumping in tape structure

So far we only considered a superconductor slab. However, when thinking of a superconductor tape, the structure is quite different, especially when considering additional copper as thermal and electrical stabiliser. Furthermore we estimate current exceeding the critical current j_c is carried in the copper stabiliser. Therefore, the current spilled over into the copper stabilisation is

$$\frac{\lambda \Delta j_c}{1-\lambda},$$

where λ is the superconductor ratio inside the tape conductor.

Therefore, the internal generated heat ΔQ_{int} (equation A.5) changes for a tape structure with electrical stabilisation into

$$\Delta Q_{int} = \mu_0 \lambda j_c \lambda \Delta j_c \frac{a^2}{3}.$$

Again we consider a linear temperature dependency of the critical current (A.1) and gain

$$\Delta Q_{int} = \frac{\mu_0 \lambda^2 j_c^2 a^2}{3(T_C - T_0)} \Delta T.$$

Considering the time τ_j to the peak value of ΔT, the overall heat balance equation becomes

$$\Delta Q_{ext} + \frac{\mu_0 \lambda^2 j_c^2 a^2}{3(T_C - T_0)} \Delta T = \gamma C \Delta T + h\tau_j \frac{\Delta T}{a}.$$

Replacing τ_j and substituting a proportionally increased stabiliser resistivity $\rho/(1-\lambda)$ we gain

$$\Delta Q_{ext} + \frac{\mu_0 \lambda^2 j_c^2 a^2}{3(T_C - T_0)} \Delta T = \gamma C \Delta T + \frac{h \Delta T \, 4a^2(1\text{-}\lambda)}{a\pi^2 \rho} \mu_0.$$

The effective specific heat for a superconducting additionally stabilised tape is

$$\gamma C_{eff} = \gamma C - \frac{\mu_0 \lambda^2 j_c^2 a^2}{3(T_C - T_0)} + \frac{h \Delta T \, 4a^2(1\text{-}\lambda)}{a\pi^2 \rho} \mu_0$$

with the condition to prevent a spontaneous flux jump for a superconductor tape of

$$\beta = \frac{\mu_0 \lambda^2 j_c^2 a^2}{\gamma C (T_C - T_0)} = 3\left(1 + \frac{4ha(1\text{-}\lambda)}{\pi^2 \rho \gamma C} \mu_0\right).$$

So far, the transport current j_t was not taken into account. When considering a transport current, ΔQ_{int} changes to

$$\Delta Q_{int} = \frac{\mu_0 \lambda^2 j_c^2 a^3}{3(T_C - T_0) 2a} \Delta T \left((1-i)^3 + (1+i)^3\right) = \frac{\mu_0 \lambda^2 j_c^2 a^2}{3(T_C - T_0)}(1 + 3i^2).$$

Together with the characteristic time τ_j calculated on basis of the longest dimension

$$\tau_j = \frac{4c^2}{\pi^2 D_m} = \frac{4a^2}{\pi^2 D_m}(1+i)^2,$$

the condition for spontaneous flux jump becomes

$$\beta = \frac{\mu_0 \lambda^2 j_c^2 a^2}{\gamma C (T_C - T_0)} \leq \frac{3}{i + 3i^2}\left(1 + \frac{4ha(1\text{-}\lambda)}{\pi^2 \rho \gamma C} \mu_0 (1+i)^2\right).$$

B Designations and abbreviations

1G	first generation
2G	second generation
3D	three dimensional
AC	alternating current
Ag	silver
AMSC	American Superconductor
Ba	barium
Bi	bismuth
CORC	Conductor on Round Core
CS	central solenoid
Cu	copper
DEMO	DEMOnstration Power Plant
Dy	dysprosium
ELM	edge-localized mode
FBI	tensile force (F) - magnetic field (B) - sample current (I)
FS	full-size
FWD	first wall dose
G10	a common glass-fibre-reinforced-plastic type
γ	gamma ray
^{1}H	hydrogen
^{2}H	deuterium
^{3}H	tritium
^{4}He	helium
HTS	high temperature superconductor
IBAD	ion beam assisted depositioning
In	indium
In33Bi67	solder consisting of 33% indium and 67% bismuth
In50Sn50	solder consisting of 50% indium and 50% tin
In52Sn48	solder consisting of 52% indium and 48% tin
In66Bi33	solder consisting of 66.3% indium and 33.7% bismuth
In80Pb15Ag5	solder consisting of 80% indium, 15% lead and 5% silver
ITEP	Institute for Technical Physics
ITER	International Experimental Thermonuclear Reactor
JET	Joint European Torus

KERI	Korea Electrotechnology Research Institute
KIT	Karlsruhe Institute of Technology
LHD	Large Helical Device
LN_2	liquid nitrogen
LTS	low temperature superconductor
ν	neutrino
O_2	oxygen
OFHC	oxygen-free high thermal conductivity copper
Pb	lead
PF	poloidal field
p^+	positron
RACC	Roebel Assembled Coated Conductor
*RE*BCO	rare-earth-barium-copper-oxide
SCS4050	copper stabilized *RE*BCO tapes from SuperPower, 4 mm wide
SEM	standard error of the mean
Sn	tin
Sn12Bi49Pb18In21	solder consisting of 12% tin, 49% bismuth , 18% lead and 21% indium
Sn40Pb60	solder consisting of 40% tin and 60% lead
Sn60Pb40	solder consisting of 60% tin and 40% lead
Sn60Pb38Cu2	solder consisting of 60% tin, 38% lead and 2% copper
Sn63Pb37	solder consisting of 63% tin and 37% lead
Sn93In4.6Pb2.3	solder consisting of 93% tin, 4.6% indium and 2.3% lead
Sn94Ag4	solder consisting of 94% tin and 4% silver
Sn96.5Ag3.5	solder consisting of 96.5% tin and 3.5% silver
Sn96.5Ag3Cu0.5	solder consisting of 96.5% tin, 3% silver and 0.5% copper
SS	stainless steel
STC	Stacked Tape Cable
TF	toroidal field
TSTC	Twisted Stacked Tape Cable
TVO	translated Russian abbreviation for "thermally resistant, water resistant, pressurized"
u	atomic mass unit
V_{RMS}	root mean square voltage
W 7-X	Wendelstein 7-X
Y	yttrium
YBCO	yttrium-barium-copper-oxide
Zr	zirconium

C Index of symbols

a	minor radius of the plasma torus (m)
a	half conductor width (mm)
A_c	contact area (mm^2)
A_{avg}	average pressure area (mm^2)
A_{Cu}	copper cross-section area (mm^2)
A_{eff}	effective pressure area (mm^2)
A_{tot}	total pressure area (mm^2)
A_{ss}	stainless steel cross-section area (mm^2)
α	fit parameter
α	thread angle (°)
B	magnetic field (T)
B_{c1}	critical magnetic field of Meißner phase (T)
B_{c2}	critical magnetic field of Shubnikov phase (T)
B_{cth}	thermodynamic critical magnetic field (T)
B_{irr}	irreversibility field (T)
β	ratio between thermal and magnetic pressure inside a plasma
β	friction angle (°)
β_{lim}	stability criterion
c_{helix}	ratio of helix pitch to arc length
C_{eff}	effective specific heat capacity ($J \cdot kg^{-1} \cdot K^{-1}$)
d_h	friction diameter (mm)
d_p	pitch diameter (mm)
ΔQ_{int}	dissipated heat (J)
ΔQ_{ext}	external heat (J)
E	electric field ($V \cdot m^{-1}$)
E_c	critical electric field ($V \cdot m^{-1}$)
ε	strain (%)
f_l	lift factor
F_{pre}	preload force (N)
γ	unit mass (kg)
h	heat transfer coefficient ($W \cdot m^{-2} \cdot K^{-1}$)
h	twist pitch (mm)
i	ratio of transfer current to critical current
I	current (A)

I_c	critical current (A)
j_c	critical current density ($A \cdot m^{-2}$)
j_{eng}	engineering current density ($A \cdot mm^{-2}$)
l_c	contact length (mm)
l_{sp}	single piece length (m)
λ	filling factor
m	number of superconductor tapes per sample
M	torque (Nm)
μ_f	friction coefficient
n	number of observations
n	sample size
n	number of HTS tapes per strand
n_{wind}	number of coil windings
N	number of measured quantities
p	pressure (Pa)
p_{avg}	average pressure (MPa)
p_{coil}	coil perimeter (mm)
p_{eff}	effective pressure (MPa)
p_{tot}	total pressure (MPa)
p_{sample}	pressure experienced by the entire sample (MPa)
p_{tape}	pressure experienced by a single tape (MPa)
P	power (W)
P_j	thermal power generated per joint (W)
q	safety factor of the toroidal magnet confinement
Q	fusion gain factor
r_{js}	specific joint resistance ($\Omega \cdot cm^2$)
R_0	major radius of the plasma torus (m)
R_c	contact resistance (Ω)
R_{curv}	curvature radius (mm)
R_{in}	intrinsic resistance (Ω)
R_j	joint resistance (Ω)
RRR	residual-resistance ratio
ρ	electrical resistivity ($\Omega \cdot mm^2 \cdot m^{-1}$)
$s(X_{i,k})$	experimental standard deviation
$s(\bar{X}_i)$	standard error of the mean (SEM)
σ_θ	hoop stress ($N \cdot m^{-2}$)
t	cable thickness (mm)
t	stabiliser thickness (mm)

t	time (s)
t_s	solder layer thickness (μm)
t_stack	superconductor stack thickness (mm)
t_tape	superconductor tape thickness (mm)
T	temperature (K)
T_c	critical temperature (K)
T_op	operation temperature (K)
T_R	room temperature (K)
T_s	specific soldering temperature (°C)
$u(x_i)$	type A standard uncertainty
$u(X_i)$	type B standard uncertainty
$u_\mathrm{c}(y)$	combined standard uncertainty
U	voltage (V)
w	cable width (mm)
w	tape width (mm)
w_tot	entire tape width required for one cable (mm)
X_i	measured quantity
$\bar{X}_i$	arithmetic mean of a measured quantity
y	arithmetic mean of the measurand
Y	measurand

Bibliography

[AAA+86] Abe, T. ; Aikawa, H. ; Akaoka, N. ; Akasaka, H. ; Akino, N. ; Akiyama, T. ; Ando, T. ; Anno, K. ; Aoyagi, T. ; Arai, T. ; Arakawa, K. ; Azumi, M. ; Fukuda, T. ; Furukawa, H. ; Hammamatsu, K. ; Haraguchi, T. ; Hayashi, K. ; Hiratsuka, H. ; Hirayama, T. ; Hiroki, S. ; Hiruta, K. ; Hitomi, N. ; Hosoda, R. ; Hosogane, N. ; Ichige, H. ; Iida, S. ; Iijima, T. ; Ikeda, Y. ; Isaji, N. ; Isaka, M. ; Ishihara, M. ; Itoh, Y. ; Kaneko, M. ; Kawamata, Y. ; Kawasaki, K. ; Kikuchi, M. ; Kimura, T. ; Kishimoto, H. ; Kitahara, K. ; Kitsunezaki, A. ; Kodama, K. ; Koide, Y. ; Koike, T. ; Kondo, I. ; Konoshima, S. ; Kubo, H. ; Kunieda, S. ; Kurihara, K. ; Kuroda, T. ; Maeda, H. ; Maeno, M. ; Matsukawa, M. ; Matsukawa, T. ; Matsuo, M. ; Miya, N. ; Miyachi, K. ; Miyo, Y. ; Mizuno, M. ; Murakami, Y. ; Mutoh, M. ; Nagami, M. ; Nagashima, A. ; Nagashima, K. ; Nagaya, S. ; Nakamura, H. ; Nakamura, Y. ; Nemoto, M. ; Neyatani, Y. ; Niikura, S. ; Ninomiya, H. ; Nishitani, T. ; Nishiyama, T. ; Nomata, H. ; Noshiroya, S. ; Ogiwara, N. ; Ohasa, K. ; Ohkubo, M. ; Ohmori, K. ; Ohmori, S. ; Ohmori, Y. ; Ohsato, Y. ; Ohshima, T. ; Ohta, M. ; Otsu, K. ; Oikawa, A. ; Ozeki, T. ; Sakasai, A. ; Sakata, S. ; Sato, M. ; Seimiya, M. ; Seki, S. ; Shiho, M. ; Shitomi, M. ; Shimada, R. ; Shimizu, K. ; Shimizu, M. ; Shimomura, Y. ; Shinozaki, S. ; Shirai, H. ; Sugie, T. ; Sunaoshi, H. ; Suzuki, K. ; Suzuki, M. ; Suzuki, S. ; Suzuki, Y. ; Tahira, S. ; Takahashi, M. ; Takahashi, S. ; Takahashi, T. ; Takatsu, H. ; Takayasu, Y. ; Takeda, S. ; Takeuchi, H. ; Takizuka, T. ; Tamura, S. ; Tanaka, E. ; Tanaka, T. ; Tani, K. ; Terakado, T. ; Tobita, K. ; Tokutake, T. ; Totsuka, T. ; Toyoshima, N. ; Tsugita, T. ; Tsuji, S. ; Tsukahara, Y. ; Tsuneoka, M. ; Ujiie, K. ; Urakawa, H. ; Uramoto, Y. ; Ushigusa, K. ; Yagyu, J. ; Yamada, K. ; Yamamoto, M. ; Yamashita, O. ; Yamashita, Y. ; Yano, K. ; Yokomizo, H. ; Yonekawa, I. ; Yoshida, H. ; Yoshikawa, M. ; Yoshino, R. : Initial experiments in JT-60. In: *Plasma Physics and Controlled Fusion* 28 (1986), Nr. 1A, 165–178. `http://dx.doi.org/10.1088/0741-3335/28/1A/015`. – DOI 10.1088/0741–3335/28/1A/015. – ISSN 0741–3335

[ABS02] Aymar, R. ; Barabaschi, P. ; Shimomura, Y. : The ITER design. In: *Plasma Physics and Controlled Fusion* 44 (2002), Nr. 5, S. 519–565. `http://dx.doi.`

org/10.1088/0741-3335/44/5/304. – DOI 10.1088/0741–3335/44/5/304. – ISBN 0741–3335

[ADC+15] AUGIERI, A. ; DE MARZI, G. ; CELENTANO, G. ; MUZZI, L. ; TOMASSETTI, G. ; RIZZO, F. ; ANEMONA, A. ; BRAGAGNI, A. ; SERI, M. ; BAYER, C. M. ; BAGRETS, N. ; CORTE, A. della: Electrical Characterization of ENEA High Temperature Superconducting Cable. In: *IEEE Transactions on Applied Superconductivity* 25 (2015), Nr. 3, 1–4. http://dx.doi.org/10.1109/TASC.2014.2364391. – DOI 10.1109/TASC.2014.2364391. – ISSN 1051–8223, 1558–2515

[Agi03] AGILENT: *Agilent 34420A Manual.* http://cp.literature.agilent.com/litweb/pdf/34420-90010.pdf. Version: 2003

[Ame15] AMELICHEV, V. : *SuperOx 2G-HTS wire: production, applications and outlook.* https://elenia.tu-bs.de/fileadmin/content/sls/8sls/SLS_Mi_7_Amelichev.pdf. Version: 2015

[AMS] AMSC: *Guidelines for Hand-Assembled Splicing of Amperium® Wire.* www.amsc.com/library/AMPSPLICING_AN_A4_0112.pdf

[Azi12] AZIZOV, E. A.: Tokamaks: from A D Sakharov to the present (the 60-year history of tokamaks). In: *Uspekhi Fizicheskih Nauk* 182 (2012), Nr. 2, S. 202. http://dx.doi.org/10.3367/UFNr.0182.201202j.0202. – DOI 10.3367/UFNr.0182.201202j.0202. – ISSN 0042–1294

[BAC+15] BAGRETS, N. ; AUGIERI, A. ; CELENTANO, G. ; TOMASSETTI, G. ; WEISS, K.-P. ; DELLA CORTE, A. : Investigation of ReBCO conductor tape joints for superconducting applications. In: *IEEE Transactions on Applied Superconductivity* 25 (2015), Nr. 3, S. 1. http://dx.doi.org/10.1109/TASC.2014.2373055. – DOI 10.1109/TASC.2014.2373055. – ISSN 10518223

[Bal15] BALSHAW, N. : *All-the-World's Tokamaks.* http://www.tokamak.info/. Version: 2015

[Bar13] BARTH, C. : *High Temperature Superconductor Cable Concepts for Fusion Magnets*, Diss., 2013. http://dx.doi.org/10.5445/KSP/1000035747. – DOI 10.5445/KSP/1000035747. – 232 S

[Bar15] BARTH, C. : *Commercial coated conductors: electrical, mechanical and thermophysical properties.* 2015

[Bay13] BAYER, C. M.: *Low resistive soldering joints for HTS tapes.* 2013

[BBC+08] BRUZZONE, P. ; BAGNASCO, M. ; CALVI, M. ; CAU, F. ; CIAZYNSKI, D. ; CORTE, A. della ; DI ZENOBIO, A. ; MUZZI, L. ; NIJHUIS, A. ; SALPIETRO, E. ; SAVOLDI RICHARD, L. ; TURTU, S. ; VOSTNER, A. ; WESCHE, R. ; ZANINO, R. : Test Results of Two European ITER TF Conductor Samples in SULTAN. In: *IEEE Transactions on Applied Superconductivity* 18 (2008), jun, Nr. 2, 1088–1091. `http://dx.doi.org/10.1109/TASC.2008.922268`. – DOI 10.1109/TASC.2008.922268. – ISSN 1051–8223

[BBD11] BRESCHI, M. ; BESSETTE, D. ; DEVRED, A. : Evaluation of Effective Strain and n-Value of ITER TF Conductor Samples. In: *IEEE Transactions on Applied Superconductivity* 21 (2011), jun, Nr. 3, 1969–1973. `http://dx.doi.org/10.1109/TASC.2010.2090850`. – DOI 10.1109/TASC.2010.2090850. – ISSN 1051–8223

[BBW+13] BARTH, C. ; BAGRETS, N. ; WEISS, K.-P. ; BAYER, C. M. ; BAST, T. : Degradation free epoxy impregnation of REBCO coils and cables. In: *Superconductor Science and Technology* 26 (2013), Nr. 5, 055007. `http://dx.doi.org/10.1088/0953-2048/26/5/055007`. – DOI 10.1088/0953–2048/26/5/055007. – ISSN 0953–2048

[BBW14] BAGRETS, N. ; BARTH, C. ; WEISS, K.-P. : Low Temperature Thermal and Thermo-Mechanical Properties of Soft Solders for Superconducting Applications. In: *Ieee Transactions on Applied Superconductivity* 24 (2014), Nr. 3, 1–3. `http://dx.doi.org/10.1109/TASC.2013.2283869`. – DOI 10.1109/TASC.2013.2283869. – ISBN 1051–8223

[Bea13] BEARDMORE, R. : *Coefficients of friction.* `http://www.roymech.co.uk/Useful_Tables/Tribology/co_of_frict.htm`. Version: 2013

[Bet39] BETHE, H. A.: Energy production in stars. In: *Physical Review* 55 (1939), S. 23

[BGB+16] BAYER, C. M. ; GADE, P. V. ; BARTH, C. ; PREUSS, A. ; JUNG, A. ; WEISS, K. P.: Mechanical reinforcement for RACC cables in high magnetic background fields. In: *Superconductor Science and Technology* 29 (2016), Nr. 2, 25007. `http://dx.doi.org/10.1088/0953-2048/29/2/025007`. – DOI 10.1088/0953–2048/29/2/025007. – ISSN 0953–2048

[BK13] BUCKEL, W. ; KLEINER, R. : *Supraleitung : Grundlagen und Anwendungen.* 7. 2013. – 508 S. – ISBN 978–3–527–66869–4

[BM08] BESSETTE, D. ; MITCHELL, N. : Review of the results of the ITER toroidal field conductor R&D and qualification. In: *IEEE Transactions on Applied Superconductivity* Bd. 18, 2008. – ISSN 10518223, S. 1109–1113

[BMS15] BARTH, C. ; MONDONICO, G. ; SENATORE, C. : Electro-mechanical properties of REBCO coated conductors from various industrial manufacturers at 77 K, self-field and 4.2 K, 19 T. In: *Superconductor Science and Technology* 28 (2015), Nr. 4, 045011. `http://dx.doi.org/10.1088/0953-2048/28/4/045011`. – DOI 10.1088/0953–2048/28/4/045011. – ISSN 0953–2048

[BOB+13] BALDAN, C. A. ; OLIVEIRA, U. R. ; BERNARDES, A. A. ; OLIVEIRA, V. P. ; SHIGUE, C. Y. ; RUPPERT, E. : Electrical and superconducting properties in lap joints for YBCO tapes. In: *Journal of Superconductivity and Novel Magnetism* 26 (2013), Nr. 5, S. 2089–2092. `http://dx.doi.org/10.1007/s10948-012-1905-6`. – DOI 10.1007/s10948–012–1905–6. – ISSN 15571947

[BOSF09] BALDAN, C. A. ; OLIVEIRA, U. R. ; SHIGUE, C. Y. ; FILHO, E. R.: Evaluation of Electrical Properties of Lap Joints for BSCCO and YBCO Tapes. In: *IEEE Transactions on Applied Superconductivity* 19 (2009), Nr. 3, S. 2831–2834. `http://dx.doi.org/10.1109/TASC.2009.2017701`. – DOI 10.1109/TASC.2009.2017701. – ISSN 10518223

[Bra] BRANDT, C. : *Grafik Wendelstein 7X*. `http://www.ipp.mpg.de/`

[Bru04] BRUZZONE, P. : The index n of the voltage-current curve, in the characterization and specification of technical superconductors. In: *Physica C: Superconductivity and its Applications* Bd. 401, 2004. – ISSN 09214534, 7–14

[Bru13] BRUKER: *YBCO Coated Conductors*. `https://www.bruker.com/fileadmin/user_upload/8-PDF-Docs/BEST/DataSheets/HTS.pdf`. Version: 2013

[BUWB15] BYKOVSKY, N. ; UGLIETTI, D. ; WESCHE, R. ; BRUZZONE, P. : Strain Management in HTS High Current Cables. In: *IEEE Transactions on Applied Superconductivity* 25 (2015), jun, Nr. 3, 1–4. `http://dx.doi.org/10.1109/TASC.2014.2360041`. – DOI 10.1109/TASC.2014.2360041. – ISSN 1051–8223

[Cel] CELENTANO, G. : *SuperOx 2G HTS wire J(T-B)*. `http://www.superox.ru/upload/SuperOx2GHTSwireJ_T-B.xls`

[Cel13] CELENTANO, G. : *Development and testing of HTS joint techniques - Final Meeting*. 2013

[CF12] CELENTANO, G. ; FABBRI, F. : *Design Assessment Studies WP12-DAS01 Superconducting Magnets Interim report*. 2012

[CJK+11] CHANG, K. S. ; JO, H. C. ; KIM, Y. J. ; CHEOLAHN, M. ; KO, T. K.: An experimental study on the joint methods between double pancake coils using YBCO coated conductors. In: *IEEE Transactions on Applied Superconductivity* 21 (2011), jun, Nr. 3 PART 3, 3005–3008. `http://dx.doi.org/10.1109/TASC.2010.2092734`. – DOI 10.1109/TASC.2010.2092734. – ISSN 10518223

[CKP+08] CHANG, K. S. ; KIM, H. ; PARK, D. K. ; KO, T. K. ; AHN, M. C. ; HA, D. H. ; SONG, J. B. ; LEE, S. J. ; KIM, H. M. ; LEE, H. : Joint characteristics of YBCO coated conductor by removing a metallic stabilizer. In: *IEEE Transactions on Applied Superconductivity* 18 (2008), jun, Nr. 2, 1220–1223. `http://dx.doi.org/10.1109/TASC.2008.920811`. – DOI 10.1109/TASC.2008.920811. – ISSN 10518223

[Cop] COPPER DEVELOPMENT ASSOCIATION INC.: *Application Data Sheet: Mechanical Properties of Copper and Copper Alloys at Low Temperatures*. `http://www.copper.org/resources/properties/144_8/`

[CPY+09] CHANG, K. S. ; PARK, D. K. ; YANG, S. E. ; KIM, Y. J. ; NA, J. B. ; KWON, N. Y. ; LEE, H. ; KIM, H. M. ; KO, T. K.: Repetitive over-current characteristics of the joints between the YBCO coated conductor. In: *IEEE Transactions on Applied Superconductivity* 19 (2009), jun, Nr. 3, 2419–2422. `http://dx.doi.org/10.1109/TASC.2009.2018201`. – DOI 10.1109/TASC.2009.2018201. – ISSN 10518223

[CPY+10] CHANG, K. S. ; PARK, D. K. ; YANG, S. E. ; JO, H. C. ; KIM, H. J. ; YOON, Y. S. ; KIM, H. S. ; LEE, H. ; KO, T. K.: Experimental analysis of a splice method between YBCO coated conductors on various bending diameters. In: *IEEE Transactions on Applied Superconductivity* 20 (2010), jun, Nr. 3, 1577–1580. `http://dx.doi.org/10.1109/TASC.2010.2042945`. – DOI 10.1109/TASC.2010.2042945. – ISBN 1051–8223

[Cra61] CRAIG, H. : Standard for Reporting Concentrations of Deuterium and Oxygen-18 in Natural Waters. In: *Science (New York, N.Y.)* 133 (1961), Nr. 3467, 1833–4.

http://dx.doi.org/10.1126/science.133.3467.1833. – DOI 10.1126/science.133.3467.1833. – ISBN 00368075

[Cre] CREDIT © ITER ORGANIZATION: *A poster of the Tokamak, "sans plasma"*. https://www.iter.org/

[DCA+15] DE MARZI, G. ; CELENTANO, G. ; AUGIERI, A. ; MUZZI, L. ; TOMASSETTI, G. ; DELLA CORTE, A. : Characterization of the critical current capabilities of commercial REBCO coated conductors for an HTS cable-in-conduit conductor. In: *IEEE Transactions on Applied Superconductivity* 25 (2015), Nr. 3, S. 3–6. http://dx.doi.org/10.1109/TASC.2014.2373651. – DOI 10.1109/TASC.2014.2373651. – ISSN 10518223

[Deu11] DEUTSCHE PHYSIKALISCHE GESELLSCHAFT: *Magnetisch eingeschlossene Fusionsplasmen*. http://www.dpg-physik.de/dpg/gliederung/fv/p/info/magnet.html. Version: 2011

[Deu16] DEUTSCHE NANOSCHICHT GMBH BASF: *High temperature superconductor tape for commercial applications*. http://www.d-nano.com/images/pdf/4317_OnePager_HTS_Tape_RZ_el_low.pdf. Version: 2016

[DKE06] DKE DEUTSCHE KOMMISSION ELEKTROTECHNIK ELEKTRONIK INFORMATIONSTECHNIK IM DIN UND VDE: *Superconductivity - Part 3: Critical current measurement - DC critical current of Ag- and/or Ag alloy-sheathed Bi-2212 and Bi-2223 oxide superconductors*. IEC 61788. 2006

[Dol13] DOLAN, T. J.: *Lecture Notes in Energy*. Bd. 19: *Magnetic Fusion Technology*. London : Springer London, 2013. – 801 S. http://dx.doi.org/10.1007/978-1-4471-5556-0. – ISBN 978–1–4471–5555–3

[DZG+10] DUCKWORTH, R. C. ; ZHANG, Y. ; GOUGE, M. J. ; REY, C. M. ; VAN DER LAAN, D. C. ; CLICKNER, C. : Voltage distribution and mechanical strength in splice joints made from as-manufactured YBCO coated conductors. In: *AIP Conference Proceedings* 1219 (2010), Nr. 1, 370–379. http://dx.doi.org/10.1063/1.3402325. – DOI 10.1063/1.3402325. – ISBN 9780735407619

[FBB+15] FLEITER, J. ; BALLARINO, A. ; BOTTURA, L. ; GOLDACKER, W. ; KARIO, A. : Characterization of roebel cables for potential use in high-field magnets. In: *IEEE Transactions on Applied Superconductivity* 25 (2015), Nr. 3, 1–1. http://dx.doi.org/10.1109/TASC.2014.2367162. – DOI 10.1109/TASC.2014.2367162. – ISSN 10518223

[FDRW+10] Floegel-Delor, U. ; Riedel, T. ; Wippich, D. ; Goebel, B. ; Rothfeld, R. ; Werfel, F. N. ; Usoskin, A. ; Rutt, A. : Copper - HTS hybrid conductor architecture. In: *Journal of Physics: Conference Series* 234 (2010), jun, Nr. 2, 022008. http://dx.doi.org/10.1088/1742-6596/234/2/022008. – DOI 10.1088/1742–6596/234/2/022008. – ISSN 1742–6596

[FKW+14] Federici, G. ; Kemp, R. ; Ward, D. ; Bachmann, C. ; Franke, T. ; Gonzalez, S. ; Lowry, C. ; Gadomska, M. ; Harman, J. ; Meszaros, B. ; Morlock, C. ; Romanelli, F. ; Wenninger, R. : Overview of EU DEMO design and R&D activities. In: *Fusion Engineering and Design* Bd. 89, Elsevier B.V., 2014. – ISBN 0920–3796, 882–889

[FQD+12] Fei Yen ; Qunxu Lin ; Donghui Jiang ; Xin Chen ; Yuanyuan Xu ; Wang, J. S. ; Wang, S. Y.: Hydrostatic Pressure Effects on the Critical Current of YBCO Coated Conductor Wire. In: *IEEE Transactions on Applied Superconductivity* 22 (2012), aug, Nr. 4, 8401103–8401103. http://dx.doi.org/10.1109/TASC.2012.2189209. – DOI 10.1109/TASC.2012.2189209. – ISSN 1051–8223

[Fra63] Frank, H. : Fünfzig Jahre Roebelstab. In: *Elektro Technik* 36 (1963), S. 709–11

[FS08] Fritz, A. H. ; Schulze, G. ; Fritz, A. H. (Hrsg.) ; Schulze, G. (Hrsg.): *Fertigungstechnik*. 8. Berlin, Heidelberg : Springer Berlin Heidelberg, 2008 (VDI-Buch). – 249 S. http://dx.doi.org/10.1007/978-3-540-76696-4. – ISBN 9783540766957

[Gad16] Gade, P. V.: *Modeling of a high current High Temperature Superconductor (HTS) conductor for use in a Toroidal Field Coil of a Fusion Power Plant*, unpublished, 2016

[GBB+14] Gade, P. V. ; Barth, C. ; Bayer, C. M. ; Fietz, W. H. ; Franza, F. ; Heller, R. ; Hesch, K. ; Weiss, K.-P. : Conceptual Design of a Toroidal Field Coil for a Fusion Power Plant Using High Temperature Superconductors. In: *IEEE Transactions on Applied Superconductivity* 24 (2014), jun, Nr. 3, 1–5. http://dx.doi.org/10.1109/TASC.2014.2299828. – DOI 10.1109/TASC.2014.2299828. – ISSN 1051–8223

[GEG+90] Grosman, A. ; Evans, T. E. ; Ghendrih, P. ; Chatelier, M. ; Mioduszewski, P. K. ; Capes, H. ; DeMichelis, C. ; Fall, T. ; Foster, A. ; Geraud, A. ; Grisolia, C. ; Guilhem, D. ; Haste, G. ; Horton, L. ;

HUTTER, T. ; LOARER, T. ; MATTIOLI, M. ; MONIER-GARBET, P. ; PECQUET, A. L. ; PEGOURIE, B. ; SAMAIN, A. ; VALLET, J. C.: Plasma edge control in Tore Supra. In: *Plasma Physics and Controlled Fusion* 32 (1990), Nr. 11, S. 1011–1027. http://dx.doi.org/10.1088/0741-3335/32/11/013. – DOI 10.1088/0741-3335/32/11/013. – ISBN 0741-3335

[GFK+09] GOLDACKER, W. ; FRANK, A. ; KUDYMOW, A. ; HELLER, R. ; KLING, A. ; TERZIEVA, S. ; SCHMIDT, C. : Improvement of superconducting properties in ROEBEL assembled coated conductors (RACC). In: *IEEE Transactions on Applied Superconductivity* Bd. 19, 2009. – ISSN 10518223, 3098–3101

[GGP+14] GOLDACKER, W. ; GRILLI, F. ; PARDO, E. ; KARIO, A. ; SCHLACHTER, S. I. ; VOJENČIAK, M. : Roebel cables from REBCO coated conductors: a one-century-old concept for the superconductivity of the future. In: *Superconductor Science and Technology* 27 (2014), sep, Nr. 9, 093001. http://dx.doi.org/10.1088/0953-2048/27/9/093001. – DOI 10.1088/0953-2048/27/9/093001. – ISSN 0953-2048

[GNK+06] GOLDACKER, W. ; NAST, R. ; KOTZYBA, G. ; SCHLACHTER, S. I. ; FRANK, A. ; RINGSDORF, B. ; SCHMIDT, C. ; KOMAREK, P. : High current DyBCO-ROEBEL Assembled Coated Conductor (RACC). In: *Journal of Physics: Conference Series* 43 (2006), jun, Nr. 1, 901–904. http://dx.doi.org/10.1088/1742-6596/43/1/220. – DOI 10.1088/1742-6596/43/1/220. – ISSN 1742-6588

[GSZV14] GRILLI, F. ; SIROIS, F. ; ZERMENO, V. M. R. ; VOJENCIAK, M. : Self-Consistent Modeling of the Ic of HTS Devices: How Accurate do Models Really Need to Be? In: *IEEE Transactions on Applied Superconductivity* 24 (2014), dec, Nr. 6, 1–8. http://dx.doi.org/10.1109/TASC.2014.2326925. – DOI 10.1109/TASC.2014.2326925. – ISSN 10518223

[HBB+08] HIRSCH, M. ; BALDZUHN, J. ; BEIDLER, C. ; BRAKEL, R. ; BURHENN, R. ; DINKLAGE, A. ; EHMLER, H. ; ENDLER, M. ; ERCKMANN, V. ; FENG, Y. ; GEIGER, J. ; GIANNONE, L. ; GRIEGER, G. ; GRIGULL, P. ; HARTFUSS, H.-J. ; HARTMANN, D. ; JAENICKE, R. ; KÖNIG, R. ; LAQUA, H. P. ; MAASSBERG, H. ; MCCORMICK, K. ; SARDEI, F. ; SPETH, E. ; STROTH, U. ; WAGNER, F. ; WELLER, A. ; WERNER, A. ; WOBIG, H. ; ZOLETNIK, S. : Major results from the stellarator Wendelstein 7-AS. In: *Plasma Physics & Controlled Fusion* 50 (2008), Nr. 5, 53001. http://dx.doi.org/10.1088/0741-3335/50/5/053001. – DOI 10.1088/0741-3335/50/5/053001. – ISBN 0741-3335

[HGF+16] HELLER, R. ; GADE, P. V. ; FIETZ, W. H. ; VOGEL, T. ; WEISS, K.-P. : Conceptual Design Improvement of a Toroidal Field Coil for EU DEMO Using High-Temperature Superconductors. In: *IEEE Transactions on Applied Superconductivity* 26 (2016), jun, Nr. 4, 1–5. http://dx.doi.org/10.1109/TASC.2016.2520662. – DOI 10.1109/TASC.2016.2520662. – ISSN 10518223

[HMB+04] HOUTTE, D. van ; MARTIN, G. ; BÉCOULET, A. ; BUCALOSSI, J. ; GIRUZZI, G. ; HOANG, G. T. ; LOARER, T. ; SAOUTIC, B. : Recent fully non-inductive operation results in Tore Supra with 6 min, 1 GJ plasma discharges. In: *Nuclear Fusion* 44 (2004), may, Nr. 5, L11–L15. http://dx.doi.org/10.1088/0029-5515/44/5/L01. – DOI 10.1088/0029–5515/44/5/L01. – ISSN 0029–5515

[HOLN04] HUSSENNETHER, V. ; OOMEN, M. ; LEGHISSA, M. ; NEUMÜLLER, H. W.: DC and AC properties of Bi-2223 cabled conductors designed for high-current applications. In: *Physica C: Superconductivity and its Applications* Bd. 401, 2004. – ISSN 09214534, 135–139

[IIA+14] ISOZAKI, K. ; IWASAKI, S. ; ANAZAWA, T. ; SASAKI, E. ; TAKAO, T. ; TSUKAMOTO, O. : Influence of Repeated Tensile Stresses on Current Transfer Characteristics of YBCO Coated Conductors With Solder Joints. In: *Ieee Transactions on Applied Superconductivity* 24 (2014), jun, Nr. 3, 1–4. http://dx.doi.org/10.1109/TASC.2013.2285100. – DOI 10.1109/TASC.2013.2285100. – ISSN 1051–8223

[Iij13] IIJIMA, Y. : High-Performance Y-based Superconducting Wire and Their Applications. Version: 2013. http://www.fujikura.co.uk/media/4338/42e_32.pdf. 2013 (Ic). – Forschungsbericht. – 117–121 S.

[Ind08] INDIUM CORPORATION OF AMERICA EUROPE AND ASIA: *Indalloy specialty alloys*. http://www.ostec-materials.ru/upload/iblock/e9a/e9aae85004a7fd1a8168520b4b742752.pdf. Version: 2008

[IOF+93] IIJIMA, Y. ; ONABE, K. ; FUTAKI, N. ; TANABE, N. ; SADAKATA, N. ; KOHNO, O. ; IKENO, Y. : In-plane texturing control of Y-Ba-Cu-O thin films on polycrystalline substrates by ion-beam-modified intermediate buffer layers. In: *IEEE Transactions on Appiled Superconductivity* 3 (1993), mar, Nr. 1, 1510–1515. http://dx.doi.org/10.1109/77.233380. – DOI 10.1109/77.233380. – ISSN 10518223

[ITE06] ITER: *Design Description Document: DDD 11 Magnet ITER_D_22HV5L v2.2.* 2006

[ITE09a] ITER: *Design Description Document - Conductors - ITER_D_2NBKXY v1.2.* 2009

[ITE09b] ITER: *Design Description Document - TF Coils and Structures - ITER_D_2MVZNX v2.2.* 2009

[ITE13] ITER: *Design Requirements and Guidelines Level 1.* 2013

[JCG08] JCGM: *Evaluation of measurement data - Guide to the expression of uncertainty in measurement.* http://www.bipm.org/utils/common/documents/jcgm/JCGM_100_2008_E.pdf. Version: sep 2008

[KAB+95] KUMAR, A. ; ABDOU, M. A. ; BARNES, C. W. ; KUGEL, H. W. ; LOUGHLIN, M. J.: Radioactivity measurements of ITER materials using the TFTR D-T neutron field. In: *Fusion Engineering and Design* 28 (1995), mar, Nr. C, 415–428. http://dx.doi.org/10.1016/0920-3796(95)90069-1. – DOI 10.1016/0920–3796(95)90069–1. – ISSN 09203796

[KBL+13] KIM, Y. ; BASCUNAN, J. ; LECREVISSE, T. ; HAHN, S. ; VOCCIO, J. ; PARK, D. K. ; IWASA, Y. : YBCO and Bi2223 coils for high field LTS/HTS NMR magnets: HTS-HTS joint resistivity. In: *IEEE Transactions on Applied Superconductivity* 23 (2013), jun, Nr. 3, 6800704–6800704. http://dx.doi.org/10.1109/TASC.2013.2243195. – DOI 10.1109/TASC.2013.2243195. – ISSN 10518223

[Kem12] KEMP, R. : *DEMO1 conservative pulsed DEMO design for PPPT studies: demo1_jul12_build_coils_2KYJLN_v1_0_2KYWWJ_v1_0.* 2012

[Kem15] KEMP, R. : *DEMO1 conservative pulsed DEMO design for PPPT studies: DEMO1_Reference_Design_-_2015_April_EU_2MDKFH_v1_0.* 2015

[KK16] KNIGHT, P. J. ; KOVARI, M. D.: *A User Guide to the PROCESS Fusion Reactor Systems Code.* http://www.ccfe.ac.uk/assets/Documents/Other/AUserGuidetothePROCESSsystemscode.pdf. Version: 2016

[KKC+09] KIM, H. S. ; KWON, N. Y. ; CHANG, K. S. ; KO, T. K. ; KIM, H. M. ; KIM, W. S. ; PARK, C. ; LEE, H. : Joint characteristics of the YBCO coated conductor (CC) by chemical etching. In: *IEEE Transactions on Applied Superconductivity* 19 (2009), jun, Nr. 3, 2835–2838. http://dx.doi.org/10.1109/TASC.2009.2018854. – DOI 10.1109/TASC.2009.2018854. – ISBN 1051–8223

[Kom95] KOMAREK, P. : *Hochstromanwendung der Supraleitung*. Teubner Verlag, 1995. – 451 S. – ISBN 3–519–03225–2

[KT98] KAYE, A. ; THE JET TEAM: Results of Recent Deuterium / Tritium Experiments in JET. 31 (1998), Nr. 40, S. 16

[Lak] LAKE SHORE CRYOTRONICS: *Appendix B: Sensor Characteristics*. http://www.lakeshore.com/Documents/LSTC_appendixB_l.pdf

[LAN36] LANDAU, L. : The Theory of Phase Transitions. In: *Nature* 138 (1936), nov, Nr. 3498, 840–841. http://dx.doi.org/10.1038/138840a0. – DOI 10.1038/138840a0. – ISBN 0028–0836

[LAP+11] LAAN, D. C. d. ; ABRAIMOV, D. ; POLYANSKII, a. a. ; LARBALESTIER, D. C. ; DOUGLAS, J. F. ; SEMERAD, R. ; BAUER, M. : Anisotropic in-plane reversible strain effect in Y0.5Gd0.5Ba2Cu3O7-δ coated conductors. In: *Superconductor Science and Technology* 24 (2011), nov, Nr. 11, 115010. http://dx.doi.org/10.1088/0953-2048/24/11/115010. – DOI 10.1088/0953–2048/24/11/115010. – ISSN 0953–2048

[Law57] LAWSON, J. D.: Some Criteria for a Power Producing Thermonuclear Reactor. In: *Proceedings of the Physical Society. Section B* 70 (1957), jan, Nr. 1, 6–10. http://dx.doi.org/10.1088/0370-1301/70/1/303. – DOI 10.1088/0370–1301/70/1/303. – ISBN 0370–1301

[LBB+08] LONG, N. J. ; BADCOCK, R. ; BECK, P. ; MULHOLL, M. ; ROSS, N. ; STAINES, M. ; SUN, H. ; HAMILTON, J. ; BUCKLEY, R. G.: Narrow strand YBCO Roebel cable for lowered AC loss. In: *Journal of Physics: Conference Series* 97 (2008), feb, 012280. http://dx.doi.org/10.1088/1742-6596/97/1/012280. – DOI 10.1088/1742–6596/97/1/012280. – ISSN 1742–6596

[LGFP01] LARBALESTIER, D. ; GUREVICH, A. ; FELDMANN, D. M. ; POLYANSKII, A. : High-Tc superconducting materials for electric power applications. In: *Nature* 414 (2001), nov, Nr. 6861, 368–377. http://dx.doi.org/10.1038/35104654. – DOI 10.1038/35104654. – ISBN 00280836 (ISSN)

[LGH12] LAAN, D. C. d. ; GOODRICH, L. F. ; HAUGAN, T. J.: High-current dc power transmission in flexible RE-Ba2Cu3O7-δ coated conductor cables. In: *Superconductor Science and Technology* 25 (2012), jan, Nr. 1, 014003. http://dx.doi.org/10.1088/0953-2048/25/1/014003. – DOI 10.1088/0953–2048/25/1/014003. – ISBN doi:10.1088/0953–2048/25/1/014003

[LHS+11] LU, J. ; HAN, K. ; SHEPPARD, W. R. ; VIOUCHKOV, Y. L. ; PICKARD, K. W. ; MARKIEWICZ, W. D.: Lap Joint Resistance of YBCO Coated Conductors. In: *IEEE Transactions on Applied Superconductivity* 21 (2011), jun, Nr. 3, 3009–3012. http://dx.doi.org/10.1109/TASC.2010.2091934. – DOI 10.1109/TASC.2010.2091934. – ISSN 1051–8223

[LLJ+13] LAAN, D. C. d. ; LU, X. ; JAGER, A. de ; BROMBERG, L. ; MINERVINI, J. ; NOYES, P. ; MILLER, G. ; WEIJERS, H. : *Conductor on Round Core magnet cables operated at high current ramp rates in background fields of up to 19 T.* http://advancedconductor.com/wp-content/uploads/2011/07/Laan-van-der-D.C.-MT23-2013.pdf. Version: 2013

[Luc11] LUCE, T. C.: Realizing steady-state tokamak operation for fusion energy. In: *Physics of Plasmas* 18 (2011), Nr. 3, 030501. http://dx.doi.org/10.1063/1.3551571. – DOI 10.1063/1.3551571. – ISBN 1070–664X

[LZL+14] LIU, W. ; ZHANG, X. ; LIU, Y. ; ZHOU, J. ; ZHOU, Y. : Lap Joint Characteristics of the YBCO Coated Conductors under Axial Tension. In: *IEEE Transactions on Applied Superconductivity* 24 (2014), dec, Nr. 6, 1–5. http://dx.doi.org/10.1109/TASC.2014.2332444. – DOI 10.1109/TASC.2014.2332444. – ISSN 10518223

[MBKF15] MESZAROS, B. ; BACHMANN, C. ; KEMP, R. ; FEDERICI, G. : Definition of the basic DEMO tokamak geometry based on systems code studies. In: *Fusion Engineering and Design* 98-99 (2015), oct, 1556–1560. http://dx.doi.org/10.1016/j.fusengdes.2015.06.097. – DOI 10.1016/j.fusengdes.2015.06.097. – ISSN 09203796

[MDL+12] MITCHELL, N. ; DEVRED, A. ; LIBEYRE, P. ; LIM, B. ; SAVARY, F. : The ITER Magnets: Design and Construction Status. In: *IEEE Transactions on Applied Superconductivity* 22 (2012), jun, Nr. 3, 4200809–4200809. http://dx.doi.org/10.1109/TASC.2011.2174560. – DOI 10.1109/TASC.2011.2174560. – ISSN 1051–8223

[Mil03] MILLER, J. R.: The NHMFL 45-T hybrid magnet system: Past, present, and future. In: *IEEE Transactions on Applied Superconductivity* Bd. 13, 2003. – ISSN 10518223, 1385–1390

[Min] MINCO PRODUCTS INC.: *Polyimide Thermofoil Heaters.* http:

//www.minco.com/~/media/WWW/ResourceLibrary/Heaters/PolyimideThermofoilHeaterTechSpec.ashx

[Moo14] MOON, S. H.: *HTS wire development and industrialization at SuNAM.* https://indico.cern.ch/event/308828/contributions/1680708/attachments/589810/811813/20140521_SuNAM_at_WAMH_DESY_vf.pdf. Version: 2014

[MW06] MARTIENSSEN, W. ; WARLIMONT, H. : *Springer handbook of condensed matter and materials data.* Bd. 1542. 2006. – 73 S. – ISBN 3540443762

[MZC+15] MIAO, Q. ; ZHU, J. M. ; CHENG, M. ; ZHANG, Z. ; LI, Z. Y. ; WANG, Y. ; SHENG, J. ; JIN, Z. ; HONG, Z. : Fabrication and Characteristic Tests of a Novel Low-Resistance Joint Structure for YBCO Coated-Conductors. In: *IEEE Transactions on Applied Superconductivity* 25 (2015), jun, Nr. 3, 1–5. http://dx.doi.org/10.1109/TASC.2014.2365694. – DOI 10.1109/TASC.2014.2365694. – ISSN 1051–8223

[Nat14] NATIONAL INSTRUMENTS: *M Series Multifunction DAQ for USB - 16-Bit, 250 kS/s, up to 80 Analog Inputs.* http://www.ni.com/datasheet/pdf/en/ds-10. Version: 2014

[Nyi06] NYILAS, A. : Transducers for sub-micron displacement measurements at cryogenic temperatures. In: *AIP Conference Proceedings* Bd. 824 I, AIP, 2006. – ISBN 0735403163, 27–34

[NZB+15] NAKASAKI, R. ; ZHANG, Y. ; BROWNSEY, P. ; SUNDARAM, A. ; HAZELTON, D. ; SAKAMOTO, H. ; FUKUSHIMA, T. : *Continuous Improvements in Performance and Quality of 2G HTS Wires Produced by IBAD-MOCVD for Coil Applications.* http://www.superpower-inc.com/system/files/2015_1018+MT24+4OrBA_03_0.pdf. Version: 2015

[OI15] OECD ; IEA: 2015 Key World Energy Statistics. Version: 2015. https://www.iea.org/publications/freepublications/publication/KeyWorld_Statistics_2015.pdf. 2015. – Forschungsbericht. – 81 S.

[OKWZ16] ONGENA, J. ; KOCH, R. ; WOLF, R. ; ZOHM, H. : Magnetic-confinement fusion. In: *Nature Physics* 12 (2016), may, Nr. May, 398–410. http://dx.doi.org/10.1038/nphys3745. – DOI 10.1038/nphys3745. – ISSN 1745–2473

[Onn11] ONNES, K. : Further experiments with liquid helium. G. On the electrical resistance of pure metals, etc. VI. On the sudden change in the rate at which the resistance of mercury disappears.Further experiments with liquid helium. G. On the electrical resistance of pure. In: *Comm. Phys. Lab. Univ. Leiden* 124c (1911), Nr. September, S. 818–821. – ISSN 0924–8323

[Oxf] OXFORD INSTRUMENTS PLC: *Magnet operator manual*

[Oxf16] OXFORD INSTRUMENTS PLC: *Personal communication.* 2016

[PAK+07] PARK, D. K. ; AHN, M. C. ; KIM, H. M. ; LEE, H. G. ; CHANG, K. S. ; LEE, S. J. ; YANG, S. E. ; KO, T. K.: Analysis of a joint method between superconducting YBCO coated conductors. In: *IEEE Transactions on Applied Superconductivity* 17 (2007), jun, Nr. 2, 3266–3269. http://dx.doi.org/10.1109/TASC.2007.899590. – DOI 10.1109/TASC.2007.899590. – ISSN 10518223

[PBL06] POLAK, M. ; BARNES, P. N. ; LEVIN, G. A.: YBCO/Ag boundary resistivity in YBCO tapes with metallic substrates. In: *Superconductor Science and Technology* 19 (2006), aug, Nr. 8, 817–820. http://dx.doi.org/10.1088/0953-2048/19/8/022. – DOI 10.1088/0953–2048/19/8/022. – ISSN 0953–2048

[PRF+69] PEACOCK, N. J. ; ROBINSON, D. C. ; FORREST, M. J. ; WILCOCK, P. D. ; SANNIKOV, V. V.: Measurement of the Electron Temperature by Thomson Scattering in Tokamak T3. In: *Nature* 224 (1969), nov, Nr. 5218, 488–490. http://dx.doi.org/10.1038/224488a0. – DOI 10.1038/224488a0. – ISSN 0028–0836

[RG84] REBUT, P. H. ; GREEN, B. J.: Status and programme of JET. In: *Plasma Physics and Controlled Fusion* 26 (1984), jan, Nr. 1A, S. 1–10. http://dx.doi.org/10.1088/0741-3335/26/1A/301. – DOI 10.1088/0741–3335/26/1A/301. – ISSN 0741–3335

[Rom12] ROMANELLI, F. : EFDA, Fusion Electricity: A roadmap to the realisation of fusion energy. Version: 2012. https://www.euro-fusion.org/wpcms/wp-content/uploads/2013/01/JG12.356-web.pdf. 2012. – Forschungsbericht. – 1–75 S. – ISBN 978–3–00–040720–8

[RVZ+04] RUPICH, M. W. ; VEREBELYI, D. T. ; ZHANG, W. ; KODENKANDATH, T. ; LI, X. : Metalorganic Deposition of YBCO Films for Second-Generation High-Temperature Superconductor Wires. In: *MRS Bulletin* 29 (2004), aug,

Nr. 08, 572–578. http://dx.doi.org/10.1557/mrs2004.163. – DOI 10.1557/mrs2004.163. – ISSN 0883–7694

[RW15] ROBERTS, K. ; WEIJERS, H. : *MagLab claims record with novel superconducting magnet.* https://nationalmaglab.org/news-events/news/maglab-claims-record-with-novel-superconducting-magnet. Version: 2015

[SBB+16] SENATORE, C. ; BARTH, C. ; BONURA, M. ; KULICH, M. ; MONDONICO, G. : Field and temperature scaling of the critical current density in commercial REBCO coated conductors. In: *Supercond. Sci. Technol.* 29 (2016), jan, Nr. 1, 1–8. http://dx.doi.org/10.1088/0953-2048/29/1/014002. – DOI 10.1088/0953–2048/29/1/014002. – ISSN 0953–2048

[Sch15] SCHNEIDER, D. : *Electromechanical test facility for the study of superconductors at cryogenic temperatures with a magnetic field*, KIT, Diss., 2015

[SD10] SHIN, H.-S. ; DEDICATORIA, M. J.: Comparison of the bending strain effect on transport property in lap-and butt-jointed coated conductor tapes. In: *IEEE Transactions on Applied Superconductivity* 20 (2010), jun, Nr. 3, 1541–1544. http://dx.doi.org/10.1109/TASC.2009.2039831. – DOI 10.1109/TASC.2009.2039831. – ISSN 10518223

[SDO10] SHIN, H.-S. ; DEDICATORIA, M. J. ; OH, S.-S. : Critical Current Degradation Behavior in Lap-Jointed Coated Conductor Tapes With IBAD Substrate Under Uniaxial Tension. In: *IEEE Transactions on Applied Superconductivity* 20 (2010), jun, Nr. 3, 1447–1450. http://dx.doi.org/10.1109/TASC.2010.2042049. – DOI 10.1109/TASC.2010.2042049. – ISSN 10518223

[SDOB09] SHIN, H.-S. ; DIZON, J. R. C. ; OH, S.-S. ; BONIFACIO, R. F.: Bending Strain Characteristics of the Transport Property in Lap-Jointed Coated Conductor Tapes. In: *IEEE Transactions on Applied Superconductivity* 19 (2009), jun, Nr. 3, 2991–2994. http://dx.doi.org/10.1109/TASC.2009.2019580. – DOI 10.1109/TASC.2009.2019580. – ISSN 10518223

[Sim16] SIMAKOV, S. : *Personal communication.* 2016

[Sip04] SIPS, A. C. C.: Advanced scenarios for ITER operation. In: *Plasma Physics and Controlled Fusion* 47 (2004), may, Nr. 5A, A19–A40. http://dx.doi.org/10.1088/0741-3335/47/5A/003. – DOI 10.1088/0741–3335/47/5A/003. – ISBN 0741–3335

[Smi00] SMITH, E. : *Mechanical Engineer's Reference Book*. 12. Butterworth-Heinemann Ltd, 2000. – 1248 S. – ISBN 978–0750642187

[SMI+04] SEKITANI, T. ; MIURA, N. ; IKEDA, S. ; MATSUDA, Y. H. ; SHIOHARA, Y. : Upper critical field for optimally-doped YBa2Cu3O7. In: *Physica B: Condensed Matter* 346-347 (2004), apr, 319–324. http://dx.doi.org/10.1016/j.physb.2004.01.098. – DOI 10.1016/j.physb.2004.01.098. – ISBN 0921–4526

[SNS+07] SUGANO, M. ; NAKAMURA, T. ; SHIKIMACHI, K. ; HIRANO, N. ; NAGAYA, S. : Stress Tolerance and Fracture Mechanism of Solder Joint of YBCO Coated Conductors. In: *IEEE Transactions on Applied Superconductivity* 17 (2007), jun, Nr. 2, 3067–3070. http://dx.doi.org/10.1109/TASC.2007.899395. – DOI 10.1109/TASC.2007.899395. – ISSN 1051–8223

[Spo15] SPONG, D. A.: 3D toroidal physics: Testing the boundaries of symmetry breaking. In: *Physics of Plasmas* 22 (2015), may, Nr. 5, 055602. http://dx.doi.org/10.1063/1.4921255. – DOI 10.1063/1.4921255. – ISSN 10897674

[SRM+08] SANTAMARIA, A. ; RUPICH, M. ; MALOZEMOFF, A. ; SUPERCONDUCTOR, A. ; MA, D. : *Scale-up of 2G HTS Wire Manufacturing at American Superconductor*. http://www.htspeerreview.com/2008/pdfs/presentations/tuesday/joint/joint_2_scale_up.pdf. Version: 2008

[Sup] SUPERCONDUCTOR TECHNOLOGIES INC.: *CONDUCTUS Benefits & Performance Advantage*. http://www.suptech.com/Benefits_and_Performance_Advantage_Oct_21.pdf

[Sup12] SUPERPOWER INC.: *SuperPower 2G HTS Wire Specifications*. http://www.superpower-inc.com/system/files/SP_2G+Wire+Spec+Sheet_for+web_2012FEC_v2_1.pdf. Version: 2012

[Sup14] SUPERPOWER INC.: *Soldering Instructions*. http://www.superpower-inc.com/system/files/SP_Soldering+Instructions_2014_v1.pdf. Version: 2014

[Sup15] SUPERPOWER INC.: 2015 Technology and Manufacturing Innovations at SuperPower. (2015), 52. http://www.superpower-inc.com/system/files/2015_SPI_New_Innovations.pdf

[TAB+12] TEDIOSI, R. ; ALESSANDRINI, M. ; BENEDUCE, C. ; SCHNEIDER, S. ; ECKERT, D. : Low Temperature and Magnetic Field Performance of Spliced Commercial YBCO CC. In: *IEEE Transactions on Applied Superconductivity* 22 (2012), jun, Nr. 3, 6600804–6600804. http://dx.doi.org/10.1109/TASC.2011.2178579. – DOI 10.1109/TASC.2011.2178579. – ISSN 1051–8223

[TB46] THOMSON, G. P. ; BLACKMAN, M. : *Improvements in or relating to gas discharge apparatus for producing thermonuclear reactions.* https://worldwide.espacenet.com/publicationDetails/biblio?FT=D&date=19590806&DB=&locale=en_EP&CC=GB&NR=817681A&KC=A&ND=1. Version: 1946

[TCBM12] TAKAYASU, M. ; CHIESA, L. ; BROMBERG, L. ; MINERVINI, J. V.: HTS Twisted Stacked-Tape Cable Conductor. In: *Superconductor Science and Technology* 25 (2012), jan, Nr. 1, 014011. http://dx.doi.org/10.1088/0953-2048/25/1/014011. – DOI 10.1088/0953–2048/25/1/014011. – ISSN 0953–2048

[THE14a] THEVA: *THEVA Pro-Line TPL 12060X AG3.* http://theva.de/wp-content/uploads/2014/07/Data-sheet-THEVA-TPL-12060X-AG3.pdf. Version: 2014

[THE14b] THEVA: *THEVA Pro-Line TPL 12060X AG3CU20.* http://theva.de/wp-content/uploads/2014/07/Data-sheet-THEVA-TPL-12060X-AG3CU20.pdf. Version: 2014

[THE14c] THEVA: *THEVA Pro-Line TPL 12100 AG3CU20.* http://theva.de/wp-content/uploads/2014/07/Data-sheet-THEVA-TPL-12100-AG3CU20.pdf. Version: 2014

[TKO+07] TAKAO, T. ; KOIZUKA, S. ; OI, K. ; ISHIYAMA, A. ; KATO, J. ; MACHI, T. ; NAKAO, K. ; SHIOHARA, Y. : Characteristics of compressive strain and superconducting property in YBCO coated conductor. In: *IEEE Transactions on Applied Superconductivity* Bd. 17, 2007. – ISSN 10518223, 3517–3519

[TMBB10] TAKAYASU, M. ; MINERVINI, J. V. ; BROMBERG, L. ; BALACHANDRAN, U. : Torsion Strain Effects on Critical Currents of HTS Superconducting Tapes. In: *AIP Conference Proceedings: Transactions of the International Cryogenic Materials Conference – ICMC* Bd. 337, 2010. – ISBN 0735407657, 337–344

[Tou16] TOURAN, N. : *Energy density calculations of nuclear fuel.* https://whatisnuclear.com/physics/energy_density_of_nuclear.html. Version: 2016

[TYT+12] TERAZAKI, Y. ; YANAGI, N. ; TOMIDA, S. ; NOGUCHI, H. ; NATSUME, K. ; MITO, T. ; ITO, S. ; HASHIZUME, H. ; SAGARA, A. : Measurement of the Joint Resistance of Large-Current YBCO Conductors. In: *Plasma and Fusion Research* 7 (2012), Nr. SPL.ISS.1, 2405027. http://dx.doi.org/10.1585/pfr.7.2405027. – DOI 10.1585/pfr.7.2405027. – ISSN 18806821

[UBS+15] UGLIETTI, D. ; BYKOVSKY, N. ; SEDLAK, K. ; STEPANOV, B. ; WESCHE, R. ; BRUZZONE, P. : Test of 60 kA coated conductor cable prototypes for fusion magnets. In: *Superconductor Science and Technology* 28 (2015), dec, Nr. 12, 124005. http://dx.doi.org/10.1088/0953-2048/28/12/124005. – DOI 10.1088/0953–2048/28/12/124005. – ISSN 0953–2048

[UWB13] UGLIETTI, D. ; WESCHE, R. ; BRUZZONE, P. : Effect of transverse load on the critical current of a coated conductor Roebel cable. In: *Superconductor Science and Technology* 26 (2013), jul, Nr. 7, 74002. http://dx.doi.org/10.1088/0953-2048/26/7/074002. – DOI 10.1088/0953–2048/26/7/074002. – ISSN 0953–2048

[VDE16] VDE VERLAG: *VDE-Bestimmungen Gruppe 3 Isolierstoffe.* http://www.vde-verlag.de/normen/gruppe3.pdf. Version: 2016

[VDL+14] VAN DER LAAN, D. C. ; DOUGLAS, F. ; LU, X. ; JAGER, A. D. ; BROMBERG, L. ; MICHAEL, P. ; MINERVINI, J. ; TROCIEWITZ, U. ; NOYES, P. ; MILLER, G. ; WEIJERS, H. : *Development of HTS Conductor on Round Core (CORC) cables for fusion applications at Advanced Conductor Technologies.* http://advancedconductor.com/wp-content/uploads/2011/07/Laan-van-der-D.C.-HTS4Fusion-2014.pdf. Version: 2014

[VGS+13] VOJENČIAK, M. ; GRILLI, F. ; STENVALL, A. ; KLING, A. ; GOLDACKER, W. : Influence of the voltage taps position on the self-field DC and AC transport characterization of HTS superconducting tapes. In: *Cryogenics* 57 (2013), oct, 189–194. http://dx.doi.org/10.1016/j.cryogenics.2013.08.001. – DOI 10.1016/j.cryogenics.2013.08.001. – ISSN 00112275

[VSW02] VINARICKY, E. ; SCHRÖDER, K.-H. ; WEISER, J. ; VINARICKY, E. (Hrsg.): *Elektrische Kontakte, Werkstoffe und Anwendungen: Grundlagen, Technologien,*

Prüfverfahren. 2. Berlin, Heidelberg : Springer Berlin Heidelberg, 2002. – 684 S. http://dx.doi.org/10.1007/978-3-642-56237-2. – ISBN 978–3–642–62698–2

[WAT+87] WU, M. K. ; ASHBURN, J. R. ; TORNG, C. J. ; HOR, P. H. ; MENG, R. L. ; GAO, L. ; HUANG, Z. J. ; WANG, Y. Q. ; CHU, C. W.: Superconductivity at 93 K in a new mixed-phase Y-Ba-Cu-O compound system at ambient pressure. In: *Physical Review Letters* 58 (1987), mar, Nr. 9, 908–910. http://dx.doi.org/10.1103/PhysRevLett.58.908. – DOI 10.1103/PhysRevLett.58.908. – ISBN 0031–9007

[WBDW16] WARMER, F. ; BEIDLER, C. D. ; DINKLAGE, A. ; WOLF, R. : From W7-X to a HELIAS fusion power plant: motivation and options for an intermediate-step burning-plasma stellarator. In: *Plasma Physics and Controlled Fusion* 58 (2016), jul, Nr. 7, 074006. http://dx.doi.org/10.1088/0741-3335/58/7/074006. – DOI 10.1088/0741–3335/58/7/074006. – ISSN 0741–3335

[WBF+16] WOLF, M. J. ; BAYER, C. M. ; FIETZ, W. H. ; HELLER, R. ; SCHLACHTER, S. I. ; WEISS, K. P.: Toward a High-Current Conductor Made of HTS CrossConductor Strands. In: *IEEE Transactions on Applied Superconductivity* 26 (2016), jun, Nr. 4, 1–4. http://dx.doi.org/10.1109/TASC.2016.2525734. – DOI 10.1109/TASC.2016.2525734. – ISSN 10518223

[Wes10] WESCHE, R. : Scaling of the Critical Current Density of Superpower RE-123. 2010. – Forschungsbericht. – 7 S.

[WFB+16] WOLF, M. J. ; FIETZ, W. H. ; BAYER, C. M. ; SCHLACHTER, S. I. ; HELLER, R. ; WEISS, K.-P. : HTS CroCo: A Stacked HTS Conductor Optimized for High Currents and Long-Length Production. In: *IEEE Transactions on Applied Superconductivity* 26 (2016), mar, Nr. 2, 19–24. http://dx.doi.org/10.1109/TASC.2016.2521323. – DOI 10.1109/TASC.2016.2521323. – ISSN 1051–8223

[Wika] WIKIPEDIA: *Chernobyl disaster*. https://en.wikipedia.org/wiki/Chernobyl_disaster

[Wikb] WIKIPEDIA: *Fukushima Daiichi nuclear disaster*. https://en.wikipedia.org/wiki/Fukushima_Daiichi_nuclear_disaster

[Wik16] WIKIPEDIA: *Helium*. https://en.wikipedia.org/wiki/Helium. Version: 2016

[Wil83] WILSON, M. N.: *Superconducting Magnets.* 1983. – ISBN 978–0198548102

[WLM+16] WEISS, J. ; LAAN, D. C. d. ; MULDER, T. ; KATE, H. T. ; GODEKE, A. ; KOLB-BOND, D. ; LARBALESTIER, D. : *Enhanced flexibility of round high-temperature superconducting CORC wires for high-field magnet applications.* https://nationalmaglab.org/images/news_events/events/mem/talks/s2_weiss.pdf. Version: 2016

[WMM+12] WALSH, R. P. ; MCRAE, D. ; MARKIEWICZ, W. D. ; LU, J. ; TOPLOSKY, V. J.: The 77-K Stress and Strain Dependence of the Critical Current of YBCO Coated Conductors and Lap Joints. In: *IEEE Transactions on Applied Superconductivity* 22 (2012), feb, Nr. 1, 8400406–8400406. http://dx.doi.org/10.1109/TASC.2011.2182349. – DOI 10.1109/TASC.2011.2182349. – ISSN 1051–8223

[WWZ10] WOLF, R. ; WARMER, F. ; ZOHM, H. : *Stellarators.* http://www.dpg-physik.de/dpg/pbh/aktuelles/pdf/Wolf.pdf. Version: 2010

[YYO11] YAMADA, Y. ; YAMADA, Y. ; OHKUMA, T. : Bending Characteristics of RE123 Wires. In: *TEION KOGAKU (Journal of the Cryogenic Society of Japan)* 46 (2011), Nr. 5, 246–253. http://dx.doi.org/10.2221/jcsj.46.246. – DOI 10.2221/jcsj.46.246. – ISSN 0389–2441

[ZDHG11] ZHANG, Y. ; DUCKWORTH, R. C. ; HA, T. T. ; GOUGE, M. J.: Solderability study of RABiTS-based YBCO coated conductors. In: *Physica C: Superconductivity and its Applications* 471 (2011), aug, Nr. 15-16, 437–443. http://dx.doi.org/10.1016/j.physc.2011.03.009. – DOI 10.1016/j.physc.2011.03.009. – ISSN 09214534

Karlsruher Schriftenreihe zur Supraleitung (ISSN 1869-1765)

Herausgeber: Prof. Dr.-Ing. M. Noe, Prof. Dr. rer. nat. M. Siegel

Die Bände sind unter www.ksp.kit.edu als PDF frei verfügbar oder als Druckausgabe bestellbar.

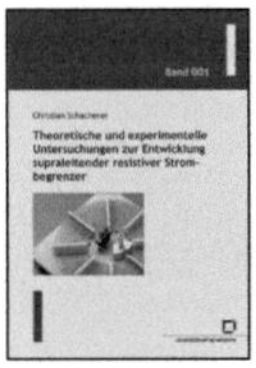

Band 001
Christian Schacherer
Theoretische und experimentelle Untersuchungen zur Entwicklung supraleitender resistiver Strombegrenzer. 2009
ISBN 978-3-86644-412-6

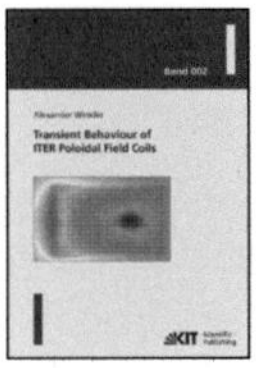

Band 002
Alexander Winkler
Transient behaviour of ITER poloidal field coils. 2011
ISBN 978-3-86644-595-6

Band 003
André Berger
Entwicklung supraleitender, strombegrenzender Transformatoren. 2011
ISBN 978-3-86644-637-3

Band 004
Christoph Kaiser
High quality Nb/Al-AlOx/Nb Josephson junctions. Technological development and macroscopic quantum experiments. 2011
ISBN 978-3-86644-651-9

Band 005
Gerd Hammer
Untersuchung der Eigenschaften von planaren Mikrowellenresonatoren für Kinetic-Inductance Detektoren bei 4,2 K. 2011
ISBN 978-3-86644-715-8

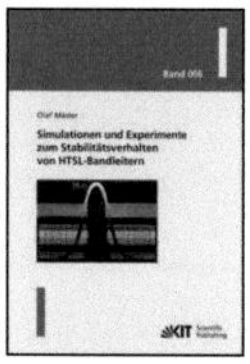

Band 006
Olaf Mäder
Simulationen und Experimente zum Stabilitätsverhalten von HTSL-Bandleitern. 2012
ISBN 978-3-86644-868-1

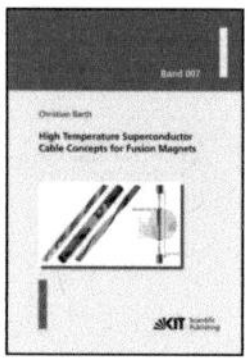

Band 007
Christian Barth
High Temperature Superconductor Cable Concepts for Fusion Magnets. 2013
ISBN 978-3-7315-0065-0

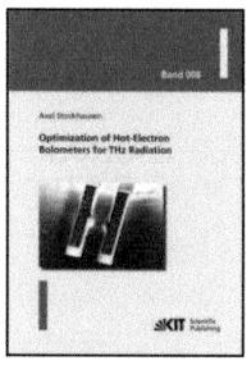

Band 008
Axel Stockhausen
Optimization of Hot-Electron Bolometers for THz Radiation. 2013
ISBN 978-3-7315-0066-7

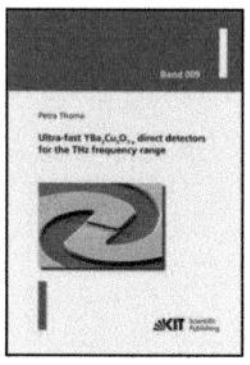

Band 009
Petra Thoma
Ultra-fast $YBa_2Cu_3O_{7-x}$ direct detectors for the THz frequency range. 2013
ISBN 978-3-7315-0070-4

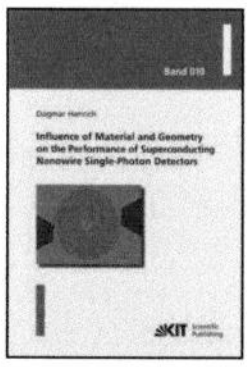

Band 010
Dagmar Henrich
Influence of Material and Geometry on the Performance of Superconducting Nanowire Single-Photon Detectors. 2013
ISBN 978-3-7315-0092-6

Band 011
Alexander Scheuring
Ultrabreitbandige Strahlungseinkopplung in THz-Detektoren. 2013
ISBN 978-3-7315-0102-2

Band 012
Markus Rösch
Development of lumped element kinetic inductance detectors for mm-wave astronomy at the IRAM 30 m telescope. 2013
ISBN 978-3-7315-0110-7

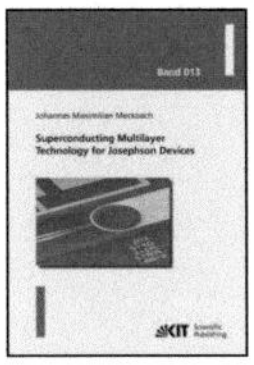

Band 013
Johannes Maximilian Meckbach
Superconducting Multilayer Technology for Josephson Devices. 2013
ISBN 978-3-7315-0122-0

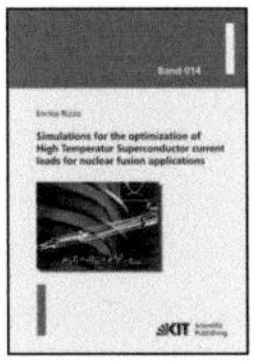

Band 014
Enrico Rizzo
Simulations for the optimization of High Temperature Superconductor current leads for nuclear fusion applications. 2014
ISBN 978-3-7315-0132-9

Band 015
Philipp Krüger
Optimisation of hysteretic losses in high-temperature superconducting wires. 2014
ISBN 978-3-7315-0185-5

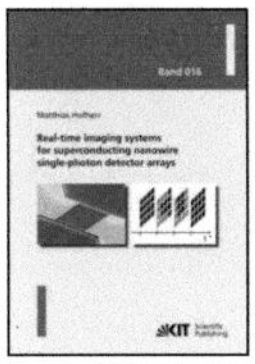

Band 016
Matthias Hofherr
Real-time imaging systems for superconducting nanowire single-photon detector arrays. 2014
ISBN 978-3-7315-0229-6

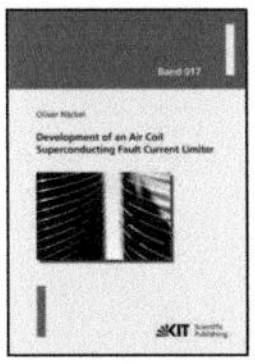

Band 017
Oliver Näckel
Development of an Air Coil Superconducting Fault Current Limiter. 2016
ISBN 978-3-7315-0526-6

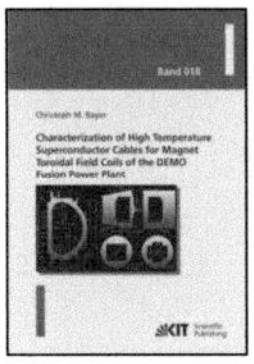

Band 018
Christoph M. Bayer
Characterization of High Temperature Superconductor Cables for Magnet Toroidal Field Coils of the DEMO Fusion Power Plant. 2017
ISBN 978-3-7315-0605-8